BRITISH ENVIRONMENTAL POLICY AND EUROPE

In the 1980s Britain was seen as the 'Dirty Man of Europe'. Since then many aspects of British policy have been profoundly Europeanised, with new environmental standards set by the European Union. The EU now specifies the cleanliness of beaches and the potability of drinking water, and traditional British approaches to environmental management and protection have been severely challenged and in many respects transformed. Adjusting Britain's procedures to a European framework has proved protracted and contentious. Tensions may now be less pronounced but over 80 per cent of British environmental legislation emanates from Brussels.

British Environmental Policy and Europe explores the consequences of recent pressures and changes, and controversial issues such as loss of sovereignty. Examining the impact of British concerns, organisations and processes on European environmental policy, and the effects of European integration on domestic environmental policies and structures, the contributors analyse a wide range of institutions and policy fields. Central and local government, quangos, pressure groups and business are examined within various fields, including pollution, land-use planning, nature conservation, water policy and waste management.

This book offers an ideal guide for all those examining contemporary environmental politics and policy in Britain or Europe. With up-to-date coverage of policy and political institutions, and wide-ranging contributions from leading observers, analysts and practitioners, a clear picture of the profound Europeanisation of British policy is presented. The editors conclude that environmental standards have been boosted but that the implementation of European Directives has incurred huge costs and Britain is still no more than a middle ranking environmental state.

Philip Lowe is Professor of Rural Economy at the University of Newcastle upon Tyne. **Stephen Ward** is a Lecturer in Politics at the University of Salford.

GLOBAL ENVIRONMENTAL CHANGE
SERIES
Edited by Jim Skea
University of Sussex

The *Global Environmental Change Series*, published in association with the ESRC Global Environmental Change Programme, emphasises the way that human aspirations, choices and everyday behaviour influence changes in the global environment. In the aftermath of UNCED and Agenda 21, this series helps crystallise the contribution of social science thinking to global change and explores the impact of global changes on the development of social sciences.

Also available in the series:

ARGUMENT IN THE GREENHOUSE
The international economics of controlling global warming
Nick Mabey, Stephen Hall, Clare Smith and Sujata Gupta

ENVIRONMENTALISM AND THE MASS MEDIA
The North–South divide
Graham Chapman, Keval Kumar, Caroline Fraser and Ivor Gaber

ENVIRONMENTAL CHANGE IN SOUTH-EAST ASIA
People, politics and sustainable development
Edited by Michael Parnwell and Raymond Bryant

POLITICS OF CLIMATE CHANGE
A European perspective
Edited by Timothy O'Riordan and Jill Jäger

THE ENVIRONMENT AND INTERNATIONAL RELATIONS
Edited by John Vogler and Mark Imber

GLOBAL WARMING AND ENERGY DEMAND
Edited by Terry Barker, Paul Enkins and Nick Johnstone

SOCIAL THEORY AND THE GLOBAL ENVIRONMENT
Edited by Michael Redclift and Ted Benton

BRITISH ENVIRONMENTAL POLICY AND EUROPE

Politics and policy in transition

Edited by Philip Lowe and Stephen Ward

Global Environmental Change Programme

London and New York

First published 1998
by Routledge
11 New Fetter Lane, London EC4P 4EE

Simultaneously published in the USA and Canada by Routledge
29 West 35th Street, New York, NY 10001

Typeset in Garamond by
RefineCatch Limited, Bungay, Suffolk
Printed and bound in Great Britain by
Biddles Ltd, Guildford and King's Lynn

British Library Cataloguing in Publication Data
A catalogue record for this book is available
from the British Library

Library of Congress Cataloguing in Publication Data
Lowe, Philip
British environmental policy and Europe / Philip Lowe and Stephen
Ward
(Global environmental change series)
Includes bibliographical references and index.
1. Environmental policy – Europe. 2. Environmental policy – Great
Britain. I. Ward, Stephen. II. Title. III. Series.
GE190.E85L68   1998
363.7'0094 – dc21   97–15808

ISBN 0–415–15500–2 (hbk)
ISBN 0–415–15501–0 (pbk)

CONTENTS

v

CONTENTS

CONTENTS

ILLUSTRATIONS

PLATES

FIGURES

ILLUSTRATIONS

TABLES

CONTRIBUTORS

Henry Buller is Maître de conférences, Département de Géographie, Université de Paris 7. He has conducted a series of comparative studies of environmental policy focusing on Britain and France, but increasingly set in a broader European context. He is co-author of *Rural Europe: Identity and Change* (1995). The research on which his chapter is based was financed by the Global Environmental Change Programme of the UK Economic and Social Research Council.

James Dixon is Senior Policy Officer of the Royal Society for the Protection of Birds. He manages a small team of specialists in agricultural and rural policy. His own responsibilities are for policy towards the Common Agricultural Policy for both the Royal Society for the Protection of Birds and BirdLife International (the global union of bird conservation organisations). He has written extensively on European agricultural and environmental issues. His chapter is based on experience in campaigning at the EU level.

Nicholas Hanley is Assistant for policy issues to the Director-General for the Environment of the European Commission. He trained as a biologist and town planner, and his career in the UK involved periods in local government, the Broads Authority and the Countryside Commission. In the latter posts, he was involved in the conflicts of agriculture and environment which led to the establishment of the first Environmentally Sensitive Areas. His first experience in Europe was with the secretariat of the Regional Committee of the Parliament. Since joining the Commission in 1989, he has co-authored the Green Paper on the Urban Environment, headed a team dealing with legislation on vehicle emissions, fuel quality and noise issues, and spent a period in the Cabinet of the previous Environment Commissioner, Yannis Paleokrassas. The views expressed in his chapter are personal to the author and do not represent the official position of the European Commission.

Andrew Jordan is a Senior Research Associate at the UK Economic and Social Research Council's Centre for Social and Economic Research on the Global Environment (CSERGE) at the University of East Anglia, where he is examining how governmental and non-governmental institutions are responding to global environmental problems in the aftermath of the 1992 Earth Summit in Rio. In the past, he has studied the implementation of various European policies in the UK and the longer-term impacts of the European Union on traditional styles and procedures of environmental protection.

Tony Long is Director of the World Wide Fund for Nature (WWF) European Policy Office based in Brussels. Prior to joining WWF, he was for five years Assistant Director of the Council for the Protection of Rural England (CPRE). He is a former Harkness Fellow of the Commonwealth Fund of New York, a former Congressional Fellow of the American Political Science Association and a Fellow of the Twenty-First Century Trust. He spent a six-month period as a visiting research fellow at the École Polytechnique in Paris. He is directing a seminar course in the College of Europe, Bruges, on shaping environmental policy in the European Union. He is a member of the editorial advisory board of the *Journal of European Public Policy*.

Philip Lowe is the Duke of Northumberland Professor of Rural Economy and Director of the Centre for Rural Economy at the University of Newcastle upon Tyne. He has long had an interest in environmental politics and has acted in an advisory capacity to various UK environmental groups and agencies. In recent years he has conducted a number of comparative studies of European environmental policy. His books include *Environmental Groups in Politics* (1983); *Rural Studies in Britain and France* (1990); *European Integration and Environmental Policy* (1993) and *Moralising the Environment* (1997).

Janice Morphet is the Chief Executive of Rutland County Council. She trained as a planner and has worked for a number of local authorities in London and the South East. Her PhD on *The Role of Local Authority Chief Executives* was published by Longman in 1993. She developed her interest in the implications of the EU for UK practice when she was at Birmingham Polytechnic, primarily through the Association of European Schools of Planning (AESOP). Since then she has been an adviser on EU environment policy to the local authority associations particularly through work on the Environment Committee of CEMR (the Committee of European Municipalities and Regions) and the Committee of the Regions. She prepared a guide to the Fifth Environmental Action Programme for the Local Government Management Board and its submission to the review of the Programme.

Martin Porter is currently a consultant on environmental issues and head of research for Adamson Associates, a Brussels-based consultancy specialising in the field of European public affairs. Prior to this he was at Bath University, where he conducted the research for his doctoral thesis on the policy networks operating around EU environmental policy issues, with special reference to the Packaging and Packaging Waste Directive. Since the award of his doctorate in 1995, he has published several articles on the subject of EU environmental policy.

Fiona Reynolds is Director of the Council for the Protection of Rural England. She holds a degree and MPhil in Land Economy from Cambridge University. She has worked in the voluntary environmental movement since 1980, first as Secretary to the Council for National Parks, then as Assistant Director (Policy) for CPRE before becoming its Director in 1992. She is Vice-President of the European Environmental Bureau, and sits on the Board, where she represents the UK members of the EEB. She has worked on a range of planning, agricultural, transport and other UK environmental issues and is a recognised authority on rural policy.

Robin Sharp CB is a retired civil servant. His overall career in the civil service ran from 1966 when he joined the Ministry of Housing and Local Government as a Principal, to 1995 when he retired from the Department of the Environment as an Under-Secretary (Grade 3). Posts in which he was directly involved in European Community work were Special Assistant to the Chancellor of the Duchy of Lancaster for European Legislation on secondment to the Cabinet Office 1972, Head of Vehicle Safety Division 1972–4, Director of Rural Affairs (responsible *inter alia* for wildlife policies) 1991–4, and Director, Global Environment 1994–5. During the second half of 1995 he was part-time Special Adviser on Central and Eastern Europe for the 'Environment for Europe' Ministerial Conference in Sofia.

Jim Skea is Director of the UK Economic and Social Research Council's Global Environmental Change Programme and a Professorial Fellow in the Environment Programme at the Science Policy Research Unit (SPRU), University of Sussex. His main interests are in environmental issues and technical change with particular reference to the energy industries. Much of his work has focused on the problems of acid rain and climate change. He participates in the Intergovernmental Panel on Climate Change. From 1993 to 1995 he held an ESRC/British Gas Fellowship in Clean Technology. He has recently conducted work assessing the implementation of integrated pollution in the UK.

Adrian Smith has recently received his DPhil from the Science Policy Research Unit (SPRU) at the University of Sussex, where he is a Research Fellow. He also holds an MSc from the Imperial College Centre for

Environmental Technology. His DPhil dissertation showed how business and regulators had interacted in implementing integrated pollution control in the chemicals industry. He has also carried out consultancy work assessing the success of pollution control systems.

Edwin Thairs trained as an applied chemist and worked with a number of companies before joining the Confederation of British Industry, where he had responsibility for environment, health and safety matters, including European liaison on these matters. In 1991 he became Head of Environment at the Water Services Association. This has involved him in a great deal of European lobbying. He is Secretary of the Groundwater Working Group of EUREAU which brings together associations of water suppliers from across the EU. He also helped to set up the European Waste Water Group and is Secretary of its Technical and Economic Committee.

Neil Ward is a Lecturer in the Department of Geography at the University of Newcastle upon Tyne and was formerly a researcher at the University's Centre for Rural Economy. He has conducted a series of studies of the politics and regulation of water quality, initially with specific focus on agricultural pollution from farm wastes and pesticides. He is co-author of *Moralising the Environment: Countryside Change, Farming and Pollution* (1997). His chapter draws upon research conducted as part of a comparative project examining the implementation of European water quality directives in five different member states. The research was funded by the European Commission's DGXII under its SEER (Socio-Economic Research on the Environment) Programme.

Stephen Ward is a Lecturer in European Politics, Department of Politics and Contemporary History, University of Salford. He previously worked as research officer at the Centre for Rural Economy, University of Newcastle upon Tyne, on an ESRC-funded project examining the impact of Europeanisation on the British environmental sector, from which this book, in part, derives. He has published a number of articles on local environmental policy and politics and is currently writing a book on UK environmental politics.

Claire Waterton is a Research Fellow at the Centre for the Study of Environmental Change, University of Lancaster. Her research centres around the construction and effectiveness of environmental policy in Europe, and has included a critique of the EU's environmental information programme (CORINE) for WWF-UK. She is currently working on an ESRC-funded project, studying the roles that different forms of knowledge about nature conservation in Europe play in the policy process.

Brian Wynne is Professor of Science Studies and Research Director of the Centre for the Study of Environmental Change at the University of

Lancaster. He has conducted various studies of European risk management and environmental policies on waste management, chemical hazards
and public risk communication, and nature conservation. He was a
research leader on risk at the International Institute for Applied Systems
Analysis, and a Visiting Scientist at the European Joint Research Centre,
Italy. He is also a European Parliament representative on the Management
Board of the European Environment Agency.

PREFACE

In preparing this book on environmental politics and policy, we had two broad purposes. The first was to give a full and up-to-date account of the institutions and issues involved in UK environmental policy. The book therefore covers the range of institutions from local authorities to central government, from pressure groups to quangos, as well as the different perspectives of environmental and business and industrial lobbies, and the role of science and public information on the environment. The range of policy sub-sectors is also covered, from pollution to conservation and from waste management to land use, to provide a review of the issues and ideas in contemporary policy debates.

A profound and pervasive influence on environmental policy these days is the European Union. Our second purpose was to identify and assess the effects of European integration on British environmental politics and policy. The book therefore examines systematically the consequences for Britain of EU policies and laws in the environmental field and of Britain's involvement with EU institutions and with other member states. We hope that what we have produced will both stand as an authoritative guide and textbook to an important policy field and illuminate the effects of one of the major forces shaping contemporary politics.

From the outset there is a need to clarify our European terminology. The various chapters refer to the European (Economic) Community (EC or EEC) and the European Union (EU). The EEC (popularly referred to as the Common Market) was created by the Treaty of Rome in 1957. The Maastricht Treaty, which came into effect in 1993, created the EU incorporating both the EEC (changing its name to the European Community) and new intergovernmental procedures for foreign and security policy, home affairs and justice policy. The EU therefore includes both federal elements (the EC) and intergovernmental elements. There are a number of institutions of the EC, the most important being the Council of Ministers, which brings together the relevant sectoral Ministers from the member states and is the main decision-making body; the Commission of the European Communities (abbreviated to CEC and more commonly known as the

European Commission) which is the civil service of the EC; and the European Parliament, which is essentially a review body with limited revising powers.

ACKNOWLEDGEMENTS

This book was drawn together during 1996–7. We would like to thank the Economic and Social Science Research Council and the European Studies Research Institute of the University of Salford for providing support to bring the authors together. The editing work relied on the support services of the Centre for Rural Economy at the University of Newcastle upon Tyne. We would thank Ruchelle Everton and Eileen Curry in particular. Figure 1.1 is reproduced by kind permission of Nigel Haigh; Plates 1.1, 2.1, 10.1 and 14.1 of Greenpeace; Plates 3.1 and 13.1 of CPRE; and Plate 12.1 of RSPB.

Part I

INTRODUCTION

1

BRITAIN IN EUROPE

Themes and issues in national environmental policy

Philip Lowe and Stephen Ward

'British environmental politics and policy cannot be understood without placing them in their European context.' This statement, which now seems a matter of fact, would have been novel or contentious a decade ago and was simply not envisaged as a possibility a decade or so earlier in 1974, when Britain joined what was then called the Common Market. Over the brief period since then a profound change has occurred in the basis of UK environmental policy through Britain's membership of the European Community. An agenda which had been driven principally by domestic factors and issues has been thoroughly Europeanised. The purpose of this book is to analyse this transformation, so as to comprehend as fully as possible the opening statement and its implications.

It might be argued that the very premise of the book renders redundant its perspective on national (and sub-national) environmental policy and politics. Should not the proper compass of environmental policy be Europe, and the appropriate object of investigation be the workings of European institutions? Such is the usual focus of European policy studies. However, the interlocking of national and European policy means that the implications for domestic procedures and politics are just as pertinent and valid a focus.

Domestic politics and European integration

The domestic politics perspective employed here is based on the proposition that, despite certain tendencies towards supranationalism, European Union politics and policy are still heavily shaped by the culture, agendas and actions of the individual member states. Such a perspective offers a means of understanding member states' differential approaches to, and implementation of, European policy. Without implementation, EU policy remains a set of formal legal texts and declamatory statements. As Haigh warns, 'there is a strong temptation to believe that the words in the text represent some kind of reality' (1984: 308). EU policy only comes to life in the member states and thus only has significance to the extent that it goads or galvanises national institutions, organisations and citizens to act. In that sense, our book should be regarded as a study of 'European environmental policy in one country'. More such studies would be needed to prepare a full account of European environmental policy and politics.

The following are the basic assumptions and justifications for adopting a domestic politics perspective towards European integration (derived in part from Bulmer 1983):

- national polities are the basic units in the European Union;
- each national polity has a particular set of social, economic and institutional conditions that shapes the national interest and policy positions and gives rise to distinctive national policy styles;
- European policy is only one facet of a national polity's activity and it is artificial to view it in isolation;
- formally, national governments hold the key positions in shaping European policy, through their membership of the Council of Ministers (and thus their responsibility for determining EU legislation) and through their dominant role in domestic politics (including their responsibility for implementing EU legislation);
- EU policy making is inevitably an amalgam of national agendas and policy styles.

The concept of national policy style is a useful explanatory device in addressing the relationship between European policy, national government and other domestic policy actors. It implies that within a country established policy making follows distinct patterns in terms of the organisation of policy, intra-sectoral relations and the usual types of policy outcomes. Clearly, such a perspective could be used to study European integration and national politics overall, as well as being applied to individual policy sectors such as the environment. Each such sector has its own preoccupations and dynamic but cannot be divorced from general national–European relations, and it is to these therefore that we turn first, before presenting a framework for analysing the Europeanisation of the environmental field.

BRITAIN AND THE EUROPEAN UNION: SEMI-DETACHMENT AND THE RELUCTANT EUROPEANS

Britain's relations with the EU have been variously portrayed as reluctant, sceptical and awkward. Rarely it seems has Britain viewed European integration or the development of European institutions with much enthusiasm. Successive governments have seemingly been preoccupied with attempts to protect British sovereignty and rearguard actions to prevent powers moving upwards to Brussels. Despite this lack of enthusiasm, integration has taken place, tacitly accepted by the same governments partly through the lack of any realistic alternatives in the prosecution of Britain's foreign and trade relations, other than closer European co-operation.

Difficulties in adapting to European integration are not peculiar to Britain. Indeed each member state likes to think it has faced particular challenges, and relations between individual states and the EU have varied over time and between different policy arenas. Yet Britain often appears to be the country most continuously and consistently at odds with European ideals and institutions. Its difficulties have been explained through a variety of historical, geographical and political factors, including:

- *Historical legacy.* Britain's large nineteenth-century empire, its former great-power status and its so-called special relationship with the United States, have left a legacy of perceptions and institutional links concerning Britain's traditional role in the world. Its outlook on trade, foreign affairs and defence have all been developed to a large extent outside the European sphere. The consequence has been that Britain has been slow to adapt to a world in which it is a middle-ranking European power. In the post-war period it failed to recognise the attractiveness of European integration to other countries; to have embraced the project would have underlined its own shrinking role in world affairs. The impetus to co-operate in post-war reconstruction was lacking and the ideal of a united Europe still fails to carry much positive resonance (Wallace 1995; Denman 1995).

- *Island mentality.* Britain's geographical detachment, combined with its historical legacy, has given rise to a different perspective towards, and a lack of practical experience of, European co-operation. Unlike most Continental countries, Britain has not experienced war on its domestic territory or occupation for several centuries. Its lack of land borders (until Irish partition) has meant that it has not been required to co-operate closely with its neighbours in managing the commerce and interactions between their respective territories. Thus the initial primary forces behind integration, namely the desire for peace, frontier security and transboundary co-operation, were not so compelling for the UK. Its distinct economic history as a global trading power traditionally favouring free trade has also meant that it is unsympathetic to some of the central policy planks of European co-operation, most notably the protectionist aspects of the Customs Union and the Common Agricultural Policy. Hence the perceived practical, economic or instrumental benefits of EU membership have rarely appeared clear cut in the UK.

- *National political structures and culture.* British political practice and institutions have also been viewed as impediments to full and enthusiastic participation in EU affairs. First, it is often argued that Britain's atypical adversarial party system does not prepare its politicians for the important process of coalition and consensus building in the European arena. Second, there has been a distinct lack of political leadership in promoting European ideals. Unlike most member states, the major British parties remain divided and have failed to create a bipartisan approach which could have developed support among the British public for European integration. Third, the British civil service has regarded the European Commission, in particular, as an alien bureaucratic structure operating on different principles (Christophe 1993). Combined with a lack of political leadership this has meant that British administrators have been circumspect in their participation in EU institutions.

Britain's scepticism towards the ideals of European unity and its problems of adaptation have produced a stance characterised as semi-detached: unavoidably part of Europe, but engaged only inconsistently; wanting economic benefits but without the political ramifications. George stresses that semi-detachment involves not simply a general scepticism and awkwardness but also a variability in European relations. In their study of institutional adaptation, he and his colleagues found considerable integration in some areas, particularly in policy arenas which remained technically defined and relatively unpoliticised. However, this steady administrative integration contrasted noticeably with the reluctance of the political forces to adapt. Essentially, separate institutions were on different learning curves (George 1992: 203).

George's study also noted that British policy within the EU was not necessarily as unified or consistent as was often presented. Policy was fragmented internally, with individual departments adapting differently and sometimes attempting to maintain their independence. Furthermore, central government's desire to play a gatekeeping role in European policy was increasingly being undermined by the ability of pressure groups and subnational government to develop their own links direct to European structures (Mazey and Richardson 1992a; Ward 1995).

Some sectors and organisations are thus much more oriented towards and accepting of the process of Europeanisation than others. In consequence, there has been differential progress in integrating the European dimension across and between sectors as well as considerable ambivalence and contention over European integration. The focus of this book is on environmental policy. This sector is a key one in national–European relations, one in which integration is well advanced and which therefore illustrates well the forces at work and the issues arising.

BRITISH ENVIRONMENTAL POLITICS:
TRADITIONS AND CHARACTERISTICS

To assess the consequences of Europeanisation in the environmental field, an understanding is first required of Britain's traditional approach. One effect of European integration is the merging of domestic and European policy. Previously these two were distinct, with policy towards European matters being a subset of the UK's foreign policy. The following account of Britain's traditional approach is therefore divided between domestic environmental policy and its international outlook in the environmental field.

British domestic environmental policy before Britain joined the Common Market

The British system of policy making has been viewed as one of competitive sectorisation with a variety of policy communities competing with one another for resources and priority on the governmental agenda. The general style of policy making within sectors has been described as one of consultative and negotiated consensus, with a desire by officials to avoid the imposition of policy solutions. The outcome of such a style is generally one of incremental change or even policy stasis (Richardson and Jordan 1979; Richardson 1982; Vogel 1986). Within this generalised British policy style, it has been commonly suggested that, until the mid-1980s, the environmental field exhibited the following characteristics (Richardson and Watts 1985; Lowe and Flynn 1989):

- *Low politics*. Environmentalism was not seen as crucial to the nation's performance and it did not arouse political contention. The major political parties paid little attention to environmental issues and they were not seen as matters of mainstream concern for central government. They were therefore pushed away from the political centre to devolved structures of administration. Even when the Department of Environment (DoE) was created in 1970, environmental protection was only one relatively small part of its functions. Many environmental issues remained outside its remit. The DoE was thus regarded by environmentalists as a department *of* the environment not *for* the environment.
- *Devolved fragmentation*. The organisation and implementation of environmental policies has tended to be devolved to local authorities, quangos and semi-independent inspectorates. Local government historically accumulated a range of environmental responsibilities based on the conviction that environmental issues were best dealt with locally. In the words of the DoE: 'Because the effects of pollution are usually experienced first within the confines of particular localities, one of the

principles followed by successive Governments has been that the primary responsibility for dealing with pollution problems should rest as far as is practicable with authorities operating at local or regional level, principally local authorities and the water authorities' (DoE 1976: 2). The use of quangos and inspectorates with statutory duties to implement and advise on environmental policy was also based on an ideal of scientific rationality: much environmental management and regulation required specialist technical knowledge that was not represented in the civil service and was best handled in the relatively depoliticised setting of an agency (Lowe et al. 1986).

- *Disjointed incrementalism.* Despite, or perhaps because of, the apparently broad compass and long history of British environmental legislation, stretching back at least to the mid-nineteenth century, environmental policy lacked overall coherence. The DoE characterised its development as 'essentially pragmatic. As an early industrial country, Britain has generally built up her law and administration stage by stage in response to particular problems' (DoE 1976: 6). The upshot of this largely incremental and reactive policy development was the absence of general principles to guide a more co-ordinated stance. To McCormick, British environmental policy is 'particularly prone to the kind of ad hoc improvisational and piecemeal responses that characterise the policy process generally'. This has led to 'a lack of direction' and 'a confusing medley of institutions and laws' (1991: 10).

- *Reliance on scientific and technical expertise.* The pragmatic, problem-solving approach to environmental regulation and management left considerable scope to the judgement of technical officials in interpreting policy. Legislation tended to be imprecise concerning standards and targets to be achieved, thus leaving its implementation very much to the discretion of officials. Policy development also relied a great deal on expert consensus through the work and deliberations of advisory committees, commissions and quangos.

- *Informal regulation.* An American commentator, David Vogel, writing in the early 1980s remarked that the British 'approach to controlling both air and water pollution makes almost no use of either legally defined standards or statutory deadlines' (1983: 57). He concluded 'if there is any one governing principle of British politics, it is that the British government should make every effort to avoid coercing its own citizenry' (1983: 75). The prevailing norms were administrative and accommodative rather than legalistic and adversarial. The desire of policy makers to ensure the co-operation of those affected by legislation involved extensive use of voluntary procedures and forms of self-regulation as well as considerable latitude in the drawing up of regulations to allow for negotiated compliance (Cox, Lowe and Winter 1990; Richardson, Ogus and Burrows 1983).

8

- *Close consultation with affected interests.* The consensual and technical approach to policy making, coupled with a concern for the practicality of legislation and a desire to avoid imposing solutions, meant close consultation in the development and implementation of policy with those economic and producer interests it was intended to influence. Environmental groups, particularly those with specialist expertise or involved in policy delivery, were also drawn into consultation, especially in such fields as nature conservation, landscape protection, historic pre-servation and town and country planning. They tended not to be included in the development of policy to do with the regulation of industrial processes and products. While environmental groups found it relatively easy to penetrate the DoE and its quangos, they tended to be excluded from other major government departments making decisions with profound environmental implications, including Transport, Energy, Trade and Industry, Agriculture, and the Treasury (Lowe and Goyder 1983).

Britain's internationalism in the environmental field before joining the Common Market

With Britain's strong tradition of devolved responsibility for environmental management and regulation, the orientation of environmental policy was self-consciously domestic. This, along with Britain's considerable experience over many years of tackling a wide range of environmental problems, made it difficult for many British interests and politicians to accept that the Euro-pean Community had any role at all to play in environmental policy. Much of the reaction in the 1970s and early 1980s was nationalistic, imbued with a sense of the practical wisdom and innate superiority of tried and tested British procedures. It would be wrong to imply, however, that the British outlook was incipiently parochial.

On the contrary, Britain had long played a leading role in environmental internationalism. It was home to a number of voluntary organisations with an international orientation, such as the Fauna and Flora Preservation Soci-ety (set up in 1903 as the Society for the Preservation of the Wild Fauna of the Empire), the Advisory Committee on the Pollution of the Seas (1952) and the International Waterfowl Research Bureau (1955). British organisa-tions and individuals had been actively involved in setting up and supporting the International Council for Bird Preservation in the interwar years and the International Union for the Conservation of Nature (IUCN) after the war (Boardman 1981). The British Government had long taken a leading role in seeking international action on such matters as combating marine pollution and protecting endangered wildlife (Zaide Pritchard 1987; Nicholson 1987). As global environmental concern grew in the 1960s and early 1970s, Britain again took a lead: for example, in 1961 a group of scientists, naturalists and businessmen set up the World Wildlife Fund (WWF); the London office of Friends of the Earth (FoE) was established in 1970; and the International Institute for Environment and Development was set up in 1973 as a joint US/UK initiative.

Britain's post-war leadership in international conservation politics, through such fora as IUCN and WWF, was in keeping with its post-colonial outlook, and the style and content reflected its position as a major scientific and cultural power (McCormick 1995). The orientation was decidedly international, concerned much more with African wildlife than with what might be happening on Britain's doorstep and focusing on *inter-* rather than *trans*-national issues, such as protection of migratory birds, endangered species and oil pollution from shipping.

Britain's island status meant that, unlike all the Continental states, it had no practical experience of cross-boundary environmental regulation such as the joint management of common rivers, lakes or watersheds. For states with such common resources, the link between domestic and international environmental issues and the need for concerted action was readily apparent. By the early 1970s, for example, the value of international bodies competent to study river pollution and make proposals for its control was well recognised in Europe. A number of such bodies had been established by treaties in the 1950s and 1960s, including international commissions for the Moselle, the Rhine, the Saar, Lake Constance and Lake Geneva, for certain rivers of common concern to Belgium, France and Luxembourg, and for frontier waters between the Federal Republic of Germany and the Netherlands and between Italy and Switzerland (Harris 1974). When the European Community moved into the field of water pollution, it drew upon this earlier experience – the Dangerous Substances in Water Directive (76/464), for example, was partly based on a new convention for the protection of the Rhine – which was understandably quite unfamiliar to Britain and elicited a considerable reaction.

Indeed, peculiarly for Britain, the international environmental agenda – to do with Third World development, population growth and natural resources – could seem quite detached from the domestic agenda – to do with land-use planning, landscape and nature conservation and pollution regulation. But the style of policy making was similar. In particular, international environmentalism was low politics: it was not the stuff of high diplomacy and, with little consequence for domestic policy, it was often left in the hands of non-governmental actors. Scientific issues and scientific expertise played a central role. It was also fragmented, oriented towards this or that international convention on, say, the regulation of whaling, the protection of migratory birds or the prevention of marine oil pollution.

Entry into the European Economic Community in 1974 raised, for Britain at least, a quite different international environmental agenda; one that was regionally rather than globally oriented; that was driven by economic considerations to do with trade liberalisation and market harmonisation; and that increasingly addressed issues of transnational environmental management that impinged directly on domestic policy. However, perhaps because of its unfamiliarity, but also because of its nascent position, the environmental policy dimension of the EEC seems to have occasioned hardly any interest or comment during Britain's entry, either in official circles or more widely. The issue was seen to be essentially to do with joining a trading bloc – the Common Market. Among adherents of a new type of environmentalism that was anxious about limits to growth, there was some apprehension. Writing in *The Ecologist* magazine, Brian Johnson (1971) suggested that Britain's traditionally steady but slow economic growth had left it with an essentially green and pleasant land which was threatened by the doctrine of 'continuous and balanced expansion' embodied in the Treaty of Rome. However, this was a fringe view that found no wider echoes, not even in the UK environmental movement, whose attention was divided between traditional domestic preoccupations and the expanding international agenda. It is perhaps significant that the initiative to set up the European Environmental Bureau as an umbrella for environmental groups across the EEC came from a meeting convened by US and Canadian environmental organisations caught up in the internationalism surrounding the Stockholm Conference.

THE DEVELOPMENT OF THE EC ENVIRONMENTAL POLICY SECTOR

Until the First Environmental Action Programme in 1973 the European Community lacked a specific environmental policy. The 1957 Treaty of Rome that established the EEC made no mention of environmental protection. It reflected instead the dominant concerns of post-war Europe with economic reconstruction, modernisation and improved living standards. The core agenda of the EEC centred on creating a free trade area and bolstering the economic growth of member states.

The origins of the EC environmental sector were as an offshoot of the economic objectives of the Community, emerging first of all in steps to avoid trade distortions and then responding to concerns about the side-effects of unbridled economic growth (Hildebrand 1992). A number of Directives mainly designed to ensure that different national environmental standards and regulatory procedures did not become an obstacle to free trade and business competition had been adopted in the 1960s. A genuine environmental policy only began, however, in the early 1970s. The 1972 summit of the heads of state of Community countries held in Paris declared that:

economic expansion is not an end in itself . . . it should result in an improvement in the quality of life as well as in standards of living . . . particular attention will be given to intangible values and to protecting the environment so that progress may really be put at the service of mankind.

(Johnson 1979)

The Community institutions were invited to develop an Action Programme on the Environment, and the first such programme was adopted the following year. It had as its aim 'to improve the setting and quality of life, and the surroundings and living conditions of the peoples of the Community'. This concern with the quality of life rather than environmental protection *per se* was in keeping with the overriding commitment of the Treaty of Rome to economic growth – a commitment that explicitly embraced the improvement and harmonisation of living and working conditions within the Community. Regulations and Decisions relating to the environment rested either on Article 100 of the Treaty (which provided for the approximation of laws between member states) or on Article 235 (which empowered the European Community to deal with any unforeseen circumstances that would otherwise impede its objectives).

Despite, or perhaps because of, its marginal status, EC environmental policy making went through its most active agenda-setting period during the 1970s. Much of what was done, such as the harmonisation of product regulation, was uncontroversial and passed largely unnoticed. But legislation was also brought forward for air, water and noise pollution, hazardous wastes and nature conservation. These established a number of environmental quality objectives and some broad 'framework' provisions regarding environmental management, for instance in the fields of water and waste. The Commission attempted to give strategic guidance in its First and Second Environmental Action Programmes (1973–6; 1977–81), which set down what have proved to be some of the fundamental notions of EC environmental policy.

It was not until the 1980s, though, that the environment was institutionalised as a distinct policy sector at the EC level. In 1981 a separate Environment Directorate-General (DGXI) was set up within the European Commission and it became the driving force behind a growing volume of environmental legislation. It also began to pay much more systematic attention to the implementation of Directives (Krämer 1989). The EC's environment programme generally enjoyed the enthusiastic support of the European Parliament and its active Environment Committee. In addition, with the general establishment of an integrated legal order within the Community, the rulings of the European Court strengthened step by step the Community's competence in the environmental field by overruling member states claiming that environmental Directives did not find sufficient basis in the EC Treaty (Koppen 1993). This, and continued pressure from the

Commission as well as from certain member states, culminated in the formal inclusion of distinct paragraphs on the environment in the Single European Act, the major Treaty revision of 1987 (Krämer 1987; Koppen 1988).

The Single European Act formalised and made explicit the strong Community involvement in the environmental field which had developed over the preceding fifteen years. At the same time, it potentially set a greater pace for environmental integration (Jans 1990). The new Article 130R, defining the general objectives and scope of EC environmental policy, states that 'environmental protection requirements shall be a component of the Community's other policies.' In the establishment of the internal market, the Single Act, rather than seeking to secure equal competition through the harmonisation of environmental standards, requires a 'high level of protection' (Article 100A(3)) to be achieved.

The institutionalisation of the EC environment sector was reflected in the gathering pace of policy making and legislation (Johnson and Corcelle 1989). A Third Environmental Action Programme, adopted in 1983, presented a far more coherent statement of policies and aims than the previous two programmes and placed greater emphasis on preventive action. The subsequent Action Programmes, adopted in 1987 and 1992, were ever more ambitious. By the Fifth Action Programme, formulated to cover the period 1993–2000, environmental protection had come to be recognised as fundamental to the sound development of the European Union (Lowe and Murdoch 1994).

Environmental policy making was the fastest-growing area of EU policy in the 1980s (Andersen and Eliassen 1990). The number of items of EU environmental legislation rose from about 5 per annum during the period of the First Action Programme to more than 20 per annum during the 1980s, reaching over 30 items per annum in the early 1990s (Haigh and Lanigan 1995).

How do we explain this phenomenal growth in EU environmental policy? Interrelated political, economic and geographic pressures played their part.

The environment was rising up the political agenda of most member states. Not only was popular concern growing, but the breakthrough to parliamentary representation of Green parties brought about the politicisation of environmental issues. Anxious governments sought to engage the Community. Several Green parties scored their first electoral results of any significance in European elections. The first Green MEPs were elected in 1984, and since 1989, when twenty-eight were elected, there has been a Green bloc in the European Parliament. The 1980s also saw the development of a Brussels-based environmental lobby. The expansion of EU environmental policy can thus be seen as a response to citizen concern across Europe.

Commission officials were not slow to stress and identify with the popularity of environmental policy. EC environmental policy was consistently and universally popular with citizens across member states, often in sharp contrast to many other European policy initiatives. The Commission was

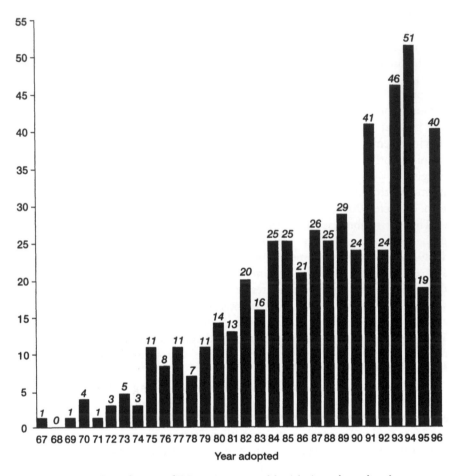

Figure 1.1 Number of items of EC environmental legislation adopted each year
Source: N. Haigh (1997) *Manual of Environmental Policy*, London: Cartermill

emboldened not only to bring forward more and more measures, but also to press for an environmental competence to be embodied in the Treaty and to propose new institutional developments such as the European Environment Agency. Members of the European Parliament (Collins and Earnshaw 1992) and the European Court of Justice (Koppen 1993) were likewise encouraged to take an activist and forward-looking approach to the pursuit of environmental protection at the Community level. Environmental policy thus emerged as a major flanking policy to the economic measures that provided the central thrust behind the movement towards European integration, and this happened at a time when the pace of integration was accelerating during the period of Jacques Delors' presidency of the European Commission from 1984.

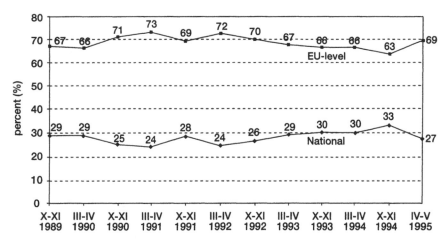

Figure 1.2 Should decisions concerning environmental protection be taken at the national level or jointly within the European Union? Surveys of European public opinion, 1989–95
Source: Europeans and the Environment (CEC 1995)

Economic factors remained a prime determinant behind the growth of the environmental sector. The planned completion of the Single European Market by the end of 1992, which became the main focus for pressing forward with market liberalisation and integration, required further harmonisation of environmental regulations and standards. Belatedly, also, the Single European Market was recognised as entailing widespread pressures on the environment that necessitated the strengthening of Community environmental policy to offset these deleterious developments (EC Task Force 1989). Other flanking policies, such as regional development, were seen to inflict their own damage through the promotion of large-scale transport and infrastructural developments, which led to additional pressures to incorporate environmental concerns into the Community's regional policy (Long 1995).

Many of the ascendant problems were transnational and global ones: acid rain, global warming, trade in endangered species, marine pollution, the shipment of toxic wastes, and so on. They revealed the limits on individual nation states in managing environmental problems on their own. The point was dramatically highlighted by a succession of disasters that showed no respect for national boundaries, including the Chernobyl nuclear accident (1986), the Rhine chemical spills, the Seveso explosion (1976), seal population crashes in the North Sea (1988–9) and the *Amoco Cadiz* oil tanker disaster (1978). The result was a stronger acceptance of the need for international environmental action.

15

In general, the 1970s and 1980s saw a dramatic growth in the number of international conventions on the environment. The development of EC environmental policy as a special type of international regime (Liefferink and Mol 1993) fitted into this wider picture. It also potentially gave EC states a strong collective voice in wider international negotiations. The late 1980s saw the Community assume a growing role as an intermediary between member states and various international fora in the environmental field, as part of its wider efforts to deploy its growing importance in world politics (Jachtenfuchs and Huber 1993). The Dublin Summit of 1989 proclaimed the obligations of the European Community 'as one of the leading collaborations of the world' to play a primary role in protecting the global environment, and it did indeed make crucial contributions to some major international decisions, such as that on the protection of the ozone layer (Jachtenfuchs 1991). As the British Government itself acknowledged, 'The "clout" that the Community carries in the wider world is greater than the sum of each Member's influence' (UK Government 1990: 37).

ADAPTATION AND INTEGRATION IN THE UK ENVIRONMENTAL SECTOR

Britain's entry into the EC in 1973 coincided with the Community's First Environmental Action Programme. Hence in this, unlike other more established policy sectors, Britain has had the opportunity to shape European environmental policy from its inception. Furthermore, Britain prided itself on being a pioneer in areas such as nature conservation and pollution regulation. Unlike some member states, it already had a considerable environmental infrastructure, legislation and institutions. With these apparent advantages, Britain might have been expected to assume a leading role in the development of European policy and to absorb this in its stride. In retrospect, however, having mature domestic structures already in place can be seen to have created difficulties. Subsequently their organisation, procedures and culture have had to be adapted to meet the requirements of European integration. This has produced both barriers and opposition to change.

Three phases of adaptation can be identified in the Europeanisation of British environmental policy.

1973–83: lack of interest and insularity

The initial decade of British membership of the EC appeared to raise few problems for the country in responding to the emerging European environmental agenda, although it did elicit a formalisation of UK policy. The general belief in government circles that EC environmental policy had very little consequence for the UK was based on a number of factors.

First, in general terms the early years of membership coincided with a period of 'Euro sclerosis', when the pace of integration slowed. The impact of the EC appeared slight across most policy sectors, and the salience of Europe declined in domestic politics after the referendum of 1975 confirming Britain's membership.

Second, in the environmental field the scale of EC legislation was small, and the British Government felt little constrained by it. The member states, and not just the UK, controlled the pace and character of EC legislation and its implementation. Measures were adopted by unanimous agreement and only after lengthy consultation, and implementation was entirely the responsibility of the member states. The UK Government adopted a particularly circumscribed view of the scope for Community action in this field – expressed by the term 'selectivity' – as set out in a 1972 memorandum 'A Policy for the European Environment':

> *selectivity*: at the Community level effort should be concentrated on work most appropriately done at that level and there should be careful choice of priorities. . . . When it comes to implementation . . . rather than pursue common legislative or administrative measures, the member states of the Communities should build severally on their existing and well-tried methods of working.
>
> (Evans 1973: 45)

In any case, the full impact of legislation agreed in the 1970s was slow to come to light, not least because of the considerable period, amounting to several years, taken to implement most European Directives.

Third, many of the issues needing to be tackled were not seen by UK officials and politicians to be relevant to the British context but as requiring action elsewhere, for example, safeguarding the cleanliness of beaches for European tourists or protecting migratory birds from slaughter or limiting industrial discharges into European waterways and enclosed seas. In the main, such perceptions were shared by the environmental lobby, who tended to see the main implications for Britain of Community membership as stemming not from EC environmental policy but from its economic and trade policies: for example, the 'juggernaut' threat posed by proposals for the upward harmonisation of permitted lorry weights or the pressure on farmland habitats from high CAP prices. Conversely, UK groups were active in supporting the extension of EC environmental policy into the field of nature conservation, but largely from a concern to strengthen international measures and to improve arrangements elsewhere in Europe rather than to change British practice.

Fourth, it was generally assumed that British practices and procedures were up to scratch and had nothing to gain in matters of standards of hygiene or amenity from Continental Europe. A sort of 'environmental chauvinism' prevailed (Caufield 1981: 416), confident in Britain's advanced status in the field of environmental affairs. A DoE guide intended to 'help those abroad who are concerned to understand the operation of pollution control here', opened with the following words:

> Great Britain has a long history of concern about the protection of the environment. . . . In consequence we are now at a comparatively advanced stage in the development and adoption of environmental protection policies.
>
> (DoE 1976: 1)

The chairman of the planning and transportation committee of the Association of County Councils was convinced that 'we in England are already on the right lines . . . Europe must learn from us. They copied us in order to have parliaments so perhaps they had better adopt our planning system' (quoted in Caufield 1981: 416). The CBI described Britain's approach as 'par excellence, one of flexibility and of consultation. It is the sort of flexible system which we should like to see more prevalent throughout the EEC' (quoted in *ibid.*).

Finally, it was felt that Britain had its own approach to environmental management which over the years had evolved a logic well fitted to the national situation. From this perspective, the very novelty of EC environmental policy and of some of the principles being proposed confirmed their lack of appropriateness to UK practice. The influential first Chairman of the Royal Commission on Environmental Pollution, Sir Eric (later Lord) Ashby, warned that the European Commission's proposals for setting limit values to potential pollutants amounted to 'a repudiation of the lessons we have learned from 160 years of our own history' (Ashby and Anderson 1981: 513). Such views echoed wider claims about the virtues of British pragmatism as against Continental abstraction, as if, through their long evolution, Britain's national procedures and practices had been tried and tested and imbued with commonsense.

Beyond these chauvinistic responses, which flourished in the absence of comparative European environmental data, what was becoming apparent was the need to explain and justify Britain's distinctive approaches to environmental management, to spell out or indeed to formulate the underlying principles. This marked the beginnings of a formalisation of environmental policy in the United Kingdom. Some of the characteristics adduced were self-consciously non-principles: the watchwords were flexibility, pragmatism, local discretion, practicability, and complexity (DoE 1976; Royal Commission on Environmental Pollution 1976). In this way, Britain's 'traditional' approach came to be defined, in reaction to the incursions of EC environmental policy making. In emphasising Britain's distinctiveness, however, the differences with the Community's approach were stressed to the point of caricature, and a coherence and commitment was claimed for British practices and procedures – particularly concerning their adherence to an environmental quality philosophy – that was not entirely warranted (Royal Commission on Environmental Pollution 1984).

Despite a general sense of the superiority of Britain's approach, there was little effort actually to proselytise it within the Community. Haigh concluded: 'The occasions when British legislation or some other initiative has shaped Community legislation are fewer than might have been expected of a country with such a well established environmental policy' (1984: 302). However, particularly in fields where it had recently overhauled its legislative and institutional structures, or where it had traditionally played an international role, Britain did take a lead. An example of the former is harmful substances: the Health and Safety at Work Act of 1974 had set up a new and comprehensive structure and procedures (including the Health and Safety Executive and Commission) and this provided a basis for Britain to take a lead in Community legislation on major industrial hazards (what eventually became the 'Seveso' Directive, 82/501/EEC) and on the testing of chemicals (the so-called 'Sixth Amendment' Directive, 79/831/EEC). Another example is waste disposal, where British practice had been comprehensively reformed by the 1974 Control of Pollution Act, which provided a model for the Community's framework Directive on waste (75/442/EEC). Examples where Britain continued to pursue its traditional international role, but now through the framework of the Community, include the Whale Regulation (348/81), which was proposed by the British Environment Secretary; and the Birds Directive (79/409/EEC), which owed a great deal to British legislation.

1983–92: defensiveness and isolation: Britain the lag state?

By the mid-1980s British insularity and practice were being increasingly challenged by the developing European environmental programme, given fresh momentum by the rising tide of green opinion across Europe. Tensions became apparent in a number of issue areas. Implementation of some legislation, such as the Bathing Water Directive and the Drinking Water Directive, proved much more costly and difficult than expected. Acid rain in Scandinavia and radioactive contamination of the Irish Sea implicated Britain as the source of major transnational pollution problems. Britain seemed more and more at odds with other member states over the direction, pace and substance of environmental policy. Finally, the reversal of long-term trends in air and water quality – including a rise in sulphur and nitrous emissions and a decline in river quality – challenged British claims for its gradualist

Plate 1.1 Greenpeace campaigners collecting sand contaminated with radioactive
discharges from Sellafield, Cumbria. The contamination of the Irish Sea by
Sellafield (then known as Windscale) contributed to Britain's image as 'the
Dirty Man of Europe'.
Source: Greenpeace/Morgan

approach to environmental improvement. The perception of Britain as a lead
nation was being reversed. Britain was now frequently depicted as an
environmental laggard, defending lax standards and blocking or watering
down efforts to strengthen European legislation. Britain's own environmental movement branded the UK 'the Dirty Man of Europe', a charge
which (fairly or otherwise) was echoed across Western Europe and Scandinavia (Rose 1990).

This unfavourable reputation rested on a number of domestic and European factors. In broad terms, Britain's general scepticism towards all things
European, the Thatcher government's combative style in European politics
and the determinedly anti-federalist rhetoric meant that the UK was viewed
as a reluctant European across most policy sectors, with the notable exception of trade liberalisation (George 1990). Britain's broad European agenda
based on a neo-liberal, deregulatory, free market approach certainly did not
fit easily with the growing regulatory approach of the Community in
environmental affairs. However, as the pace of Europeanisation increased,
so Britain, with its strong interest in the Single European Market, was
obliged to accept an ever expanding role for the EC in environmental policy
making, which other member states pressed for as the counterpart to
market liberalisation.

Britain's reputation for awkwardness and foot dragging was established through a series of clashes with European partners and the Commission: over the Directive on discharges to water of dangerous substances, acid rain and the Large Combustion Plants Directive, over its opposition to the Environmental Impact Assessment (EIA) Directive, and over its partial implementation of waste and water quality Directives (Boehmer-Christiansen and Skea 1991; Chapters 10, 13 and 14 this volume). The Royal Commission on Environmental Pollution, no less, accused the British Government of pursuing 'artificially entrenched positions' (1984: 48), of turning 'the much proclaimed virtues of flexibility . . . into a form of dogma' (48), that 'in some cases produced results which border on the ridiculous' (52) and of having 'over-protested in favour of domestic practices' (53). It warned that 'the United Kingdom must respond to the views of its European Community partners, and it should do so in a way that ensures that constructive criticism is not mistaken for obstructiveness' (53). Unfortunately, in the late 1980s tensions between the UK Government and the Commission escalated. Between 1989 and 1992, under the flamboyant Environment Commissioner, Carlo Ripa di Meana, and in a succession of high-profile cases, the Commission threatened legal action against the British Government over breaches of the EIA Directive, polluted bathing waters, the low number of air pollution monitoring stations and the shooting of 'pest' birds.

During the 1980s, a profound change occurred in the outlook of British environmental groups. As part of an increasingly oppositional stance that was ever more critical of the UK's performance, campaigners came to see the European Community as an environmental beacon – it provided them with issues, with vital environmental information (from the monitoring requirements of Directives) and with authority to challenge domestic practices and procedures. By the end of the decade the European Commission was receiving from the UK the largest number of complaints concerning infringements of environmental Directives, amounting to one-third of the total (Collins and Earnshaw 1992). Although this does not necessarily mean that the UK had the worst record, it does illustrate the new activism of British environmentalists in using European channels to pursue conflicts with their own government, and at the time they found a willing champion in DGXI. Undoubtedly, the publicity and public support attracted by such campaigns added considerably to the impression of a member state that was struggling to come to terms with European environmental policy.

In this context, and with heightened public concern over green issues, the threats of legal action from the European Commission transformed what might have been comparatively minor wranglings over the pace of implementation and scope for national interpretation of Directives into matters of high politics. The embarrassment caused elicited a furious response from

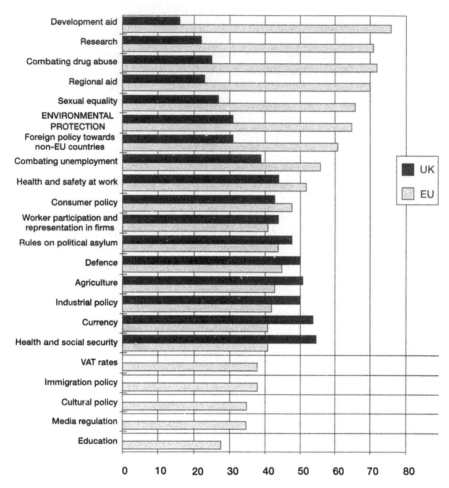

Figure 1.3 Survey of opinion in the UK concerning policy areas: should they be
decided at the national level or jointly within the European Union?
Source: Europeans and the Environment (CEC 1995)

the UK Government. In particular, a letter from the Commission demanding that work stop on the contentious Twyford Down road scheme while an
environmental impact assessment was carried out drew a retort from the
Prime Minister, John Major, allegedly threatening to block the signing of the
Maastricht Treaty and, more realistically, the proposed extension of majority
voting for European environmental legislation (*Independent* 22 October
1991).

From 1992: environmental slow-down and converging agendas?

Having reached their nadir in 1991, there was only one direction for British–EC environmental relations to go. During the following year, the high-profile conflict between the UK and the European Community over environmental policy subsided. Although Britain still faced acute difficulties coming to terms with European integration, the tone of Britain's European diplomacy altered significantly: gone was Margaret Thatcher's 'handbagging' of European leaders, to be replaced by John Major's aspiration for Britain to be 'at the heart of Europe'. Differences in the environmental sector were also less pronounced after 1992 than previously. Arguably this was the result of a convergence between European and UK agendas following a tacit recognition by elites on either side that squabbling over environmental matters must not be allowed to upset efforts to negotiate the larger and more difficult project of economic and monetary union.

Environmental convergence came about partly as a consequence of the efforts of the UK Government to refurbish environmental policy in the late 1980s and early 1990s. Initiatives such as *This Common Inheritance* (the environmental White Paper of 1990), *Sustainable Development: The UK Strategy* (1994, setting out the UK Government's response to the Rio Earth Summit) and the creation of the Environment Agency (1996), while not free of criticism, represented a concerted attempt to place environmental policy on a coherent and systematic basis. These initiatives were needed politically to overcome a crisis of public confidence in domestic environmental policy brought about largely through European pressures and environmental campaigning and manifest most clearly in the 15 per cent of the popular vote cast for the Green Party in the European elections of 1989. An aspect of this modernisation of policy was to bring the UK into step with European and international developments. Indeed, following Margaret Thatcher's 'green speech' of 27 September 1988 about the threats to mankind posed by atmospheric pollution, there was a determined effort to regain a position of leadership for the UK in international environmental diplomacy (Flynn and Lowe 1993).

On the other hand, environmental convergence was assisted by a significant slackening in the pace of European environmental integration. The volume of new environmental legislation fell and the Fifth Environmental Action Programme made slow progress. This slow-down stemmed partly from general doubts about further European integration following the difficulties experienced by several member states in gaining approval for the Maastricht Treaty, as well as a lack of enthusiasm for further environmental regulation amid economic recession and high levels of unemployment. An expected boost to European environmental policy making, from the entry of such greenish states as Sweden, Finland and Austria and from the extension of majority voting and the role of the Parliament under the Maastricht Treaty, failed to materialise (Anderson and Liefferink 1997). After the departure of Ripa Di Meana as Environment Commissioner in 1992, DGXI's internal influence also waned. The Directorate appeared to lack strong political leadership or support as environmental attention declined across Europe. There was also a change in emphasis signalled by the Fifth Environmental Action Programme involving a move away from the legislative factory approach of the 1980s, towards a programme built around mixed policy instruments and shared responsibility (Lowe and Murdoch 1995; Ward and Williams 1997; ENDS 1996a).

Traditional British governmental preoccupations with the practicality of Directives, the costs and benefits of environmental legislation, the flexibility of implementation, and the standardisation of monitoring and compliance procedures, all received more favourable and prominent attention than previously. This was partly because they resonated with the Commission's and some other member states' concern over the poor record of implementing existing environmental legislation. It was widely accepted that European environmental policy was often ignored and that some member states lacked the administrative capacity, let alone the commitment, to comply with legislation (La Spina and Sciortino 1993). Implementation and compliance became an obvious focus of effort during a period of policy consolidation and retrenchment.

The British Government was also able to exploit to its advantage the changing balance of European relations. On the one hand, it used arguments to do with subsidiarity to claw back aspects of sovereignty. The Single European Act of 1987 had addressed longstanding British concerns by making specific reference to subsidiarity in the new environment section of the treaty (Golub 1996): the Community was thereby mandated to take action relating to the environment where the objectives 'can be attained better at Community level than at the level of the individual Member States' (Article 130R(4)). The Maastricht Treaty built on and generalised this provision. Soon after the Treaty had been signed, the British Government made use of its tenure of the Council Presidency in 1992 to lead an offensive to redefine the scope of Community environmental policy in accordance with the subsidiarity principle. The need for it to justify Community action more fully led the Commission to withdraw certain proposals and initiate many fewer of them: the number of environmental proposals halved between 1992 and 1995 (Golub 1996). Subsidiarity was also behind the shift in the type of instrument adopted, away from Regulations and other detailed or binding measures and towards framework Directives, soft law and voluntary codes that maintained as much national discretion as possible. A recent assessment of the impact of the subsidiarity provision on EU environmental policy concludes that 'Britain has had the greatest success in preventing Community environmental action' (Golub 1996: 700).

While the UK Government used subsidiarity to limit EC power in relation to domestic policy, it also took and promoted opportunities to bypass the Community's federal structures in pan-European and international initiatives. For example, whereas the Government had been unco-operative towards the Commission's efforts to establish an environmental information programme – CORINE (Co-ordination of Information on the Environment) – which operated between 1985 and 1991 under DGXI, it was enthusiastic about the subsequent setting up of a European Environment Agency, separate from the Commission and under intergovernmental supervision (Chapter 7). It also promoted the formation in 1992 of the so-called IMPEL network of member state agencies to exchange information on the national enforcement of environmental regulations (see p. 99). Likewise, the UK remained an active, and often leading, player in various free-standing regional conventions or ones organised by the United Nations Economic Commission for Europe (UNECE). For example, with the prospect of enlargement of the EU to the east, Britain supported the development of joint environmental programmes through the 'Environment for Europe' initiative, by which Western governments co-operated with Central and Eastern European governments to address a pan-European agenda in the post-Communist era. After the ineffectual involvement of the Commission in the Earth Summit at Rio and in the negotiations over the reduction of greenhouse gases (Haigh 1996; Macrory and Hession 1996), the emphasis of EU involvement in international negotiations was more on co-ordination of member states' positions through the Council of Ministers. Increasingly, the Council was used to establish broad agreement on the approach to be adopted in regional and international conventions, and if necessary a common position on particular points. For the UK Government such positions were normative, but this form of intergovernmental co-ordination suited it and fitted in well with a more activist role for the UK in global environmental politics, following up the Earth Summit and such international conventions as those on climate change, the ozone layer and biodiversity.

THE IMPACT OF EUROPEAN ENVIRONMENTAL POLITICS AND POLICY

Despite the recent relative success of the British Government in limiting EC power, there has since the 1970s undoubtedly been a major shift in the locus of environmental policy making towards the European level. It is estimated that 80 per cent of UK environmental legislation has its origins in Brussels and Strasbourg (Gummer 1994). The overall impact in the UK, which is the subject of the book, can be assessed along the three dimensions of policy style, policy relations and policy substance.

Policy style: flexibility to formality?

There is little dispute that European integration has challenged Britain's traditional means of managing and administering the environment. The environmental policy style of the UK has been shaped by Europeanisation in a number of respects.

First, the profile of environmental issues has clearly been heightened by European involvement. Environmental politics has become a matter of international concern and high politics. Central government, through its European commitments, has been obliged to take environmental policy more seriously as a legitimate responsibility. Europeanisation was one of the processes which helped propel the environment up the domestic agenda in the 1980s. European debates forced central government to crystallise and formalise national policy (Flynn and Lowe 1993).

Second, Britain's fragmented, incremental, pragmatic, case-by-case approach has been challenged by European policy which often has its basis in general principles such as the precautionary approach or the polluter pays (Hill et al. 1989).

Third, Britain's flexible, administrative style based around negotiated voluntary consent, the lack of fixed national targets and the use of departmental circulars to implement policy has been progressively undermined by European legislation with its much more formal and explicit approach. European regulation has therefore brought with it greater codification, transparency and more legalistic measures inherent in civil (Roman) law and federal systems (Jordan 1993; Chapter 10).

Policy making and organisational roles and relations: destabilisation and centralisation?

The position of certain groups and institutions has been enhanced by the processes of Europeanisation, whereas others have been marginalised. This, in turn, has impacted on the way environmental policy is made and on the policy relations between environmental institutions.

One suggestion is that Europeanisation in the environmental field has opened up or destabilised some of the traditionally closed domestic policy communities such as water and agriculture. In general, the policy agenda has become much harder to control and contain within domestic networks, when the policy climate is frequently being determined elsewhere and the locus of policy making has moved to Brussels (Richardson and Maloney 1994).

Interlinked to the opening up of some policy communities is the way European legislation has allowed new policy actors to gain political access. Environmental groups, in particular, seem to have benefited. The new European tier allows groups to bypass Whitehall or appeal to the Commission as a court of redress against UK Government decisions (see Chapter 5). Furthermore, the introduction of European directives into UK law has provided groups with a great deal of new information not available under the traditionally closed and confidential regime of UK policy negotiations. Essentially the more codified nature of European environmental legislation has furnished groups with a yardstick to measure governments' progress. The European Commission, with a limited capacity and authority to monitor the implementation activities of the member states, has in turn encouraged environmental groups to develop a watchdog role to monitor the performance of their own governments (Potter and Lobley 1990; Chapter 5).

The relationships between organisations and institutions have also been altered. For example, there is considerable debate about the effects of European environmental legislation on central–local relations and responsibilities in the UK. On the one hand, Europe has been viewed as a centralising force, where statutory powers have shifted upwards through the governmental tiers. Thus the traditionally strongly devolved nature of environmental administration in the UK has been weakened. Local authorities have been stripped of responsibilities and discretionary authority. Haigh argues that this is because it is central government that is formally responsible for EU negotiations and is held responsible for any breaches of compliance. Thus the autonomy of local authorities has had to be curtailed (Haigh 1986; Haigh and Lanigan 1995).

This experience is not necessarily mirrored in other member states (Chapter 4), which suggests that it may be necessary to disentangle the effects of European integration and domestic pressures towards centralisation and decentralisation. Morphet, for example, suggests that as environmental powers have moved from Whitehall to Brussels, civil servants have compensated for this reduction in their powers by taking over functions from local government (see Chapter 8).

The reduction of statutory powers does not necessarily paint the whole picture. As with the environmental groups, local authorities have often been keen to bypass central government and foster closer ties with EU institutions, particularly the European Commission. Local authorities have tended to view the Commission as being more environmentally sympathetic, giving more of a lead, and providing resources to assist with the implementation of environmental programmes. Some commentators have begun to see in these developments a strengthening of the position of sub-central government as direct contacts, extensive networking and institutional channels of dialogue have emerged within the EU (Ward 1995; Ward and Williams 1997).

Policy substance: higher standards?

The impact of Europeanisation on the substance of UK environmental policy is the most difficult dimension to analyse, since one is making comparisons with the hypothetical scenario of what might have happened had the UK remained outside the EC.

Two contradictory viewpoints have been put forward. Lee (1992) argues that in the majority of cases the EC has made no significant difference to actual policy because the UK Government would probably have introduced measures with similar effect, though possibly in a different form. This line of argument would seem to suggest that many of Britain's difficulties lie not with the substance of policy but more with the politics and rhetoric of environmentalism and European integration.

The more common viewpoint is that Community membership has brought higher standards to the UK than would otherwise have been the case. This is based partly on the notion that the Commission in principle takes its benchmark for environmental standards from the greenest states and therefore pressurises the majority of member states to ratchet up their standards. However, this is not necessarily the end result since all legislation is subject to political bargaining and trade-offs. Environmentalists have complained that this leads to 'lowest common denominator' legislation, whereby the lag states actually set the pace of environmental change. Nevertheless, despite the bargaining and the watering down of directives most UK commentators argue that European policy has increased pressure for higher standards in the UK (Osborn 1992).

THE BOOK

In this chapter we have sought to set out a framework for analysing the Europeanisation of British environmental policy and politics. The overall aim of this book is to address the question of how effective the British Government and organisations in the environmental field have been in responding to the challenge of European integration. Individual chapters will address the following points:

- the impact of British concerns, organisations and processes on European environmental policy;
- the effect of European integration on domestic environmental politics and structures.

Here we briefly recap our argument and draw out the specific questions that the book is seeking to answer.

Britain and the EU

Britain's relations with the European Community and Union have been variously portrayed as reluctant, sceptical and awkward. Overall, the relationship has been characterised as 'semi-detached'. Some sectors and groups, though, are much more oriented towards, and accepting of, the process of Europeanisation than others. In consequence, there has been differential progress in integrating the European dimension across and between sectors, as well as considerable ambivalence and contention over European integration. *How central has the environment been to UK–EU relations compared to other sectors? Which groups and organisations have been in the forefront promoting the European dimension of environmental policy and which ones have resisted?*

The characteristics of British domestic environmental policy

The traditional style of British environmental policy would be characterised as: low politics; pragmatic; piecemeal and incremental; reliant on scientific and expert consensus; and involving close (and closed) consultation with affected interests; devolved implementation; and a preference for informal regulation. *How has this style changed in reaction or response to European integration? Can one still speak of a distinctive national policy style? What are the characteristics of current approaches to environmental policy?*

Britain was a leader in international conservation politics in the post-war world. Britain's leadership position was in keeping with its post-colonial outlook and the style and content reflected its position as a major scientific and cultural power, as well as its island status. *When and how did Britain's environmental diplomacy change following EC entry? What are the determinants of Britain's current environmental diplomacy in Europe?*

Adaptation and integration in the British environmental sector

We have suggested that Britain has experienced a long process of adjustment to the emerging European framework for environmental policy and we have identified three phases of adaptation: 1973–83: a period of lack of interest and of insularity; 1983–92: a period of defensiveness and isolation; and after 1992: a period of convergence between British and EU agendas. *How have the attitudes and tactics of central government and other policy actors towards the EC/EU changed between these periods? What were the consequences domestically and internationally of coming to be perceived as a lag state? What was the response to this perception by different policy actors? Is this label still appropriate? What examples are there where Britain has made the running in the development of EU policy?*

29

The impact of European environmental politics and policy

Since 1973, there has undoubtedly been a major shift towards Brussels in the locus of environmental policy making. The overall impact in the UK can be assessed along the three dimensions of policy style, policy relations and policy substance. *What have been the changes in procedures and principles in the different fields of environmental policy as a result of European developments? What have been the consequences for the relationships between policy actors in the environmental field? What have been the implications for standards of environmental management and protection in the UK?*

Prospects

Arguably, Britain and the EU have been involved in a process of mutual adaptation in this and other policy fields. *What lessons can be learnt from this experience and what should be the strategy for the British environmental sector in Europe for the future?*

The rest of the book is divided into three main parts. Part II focuses on the key axis between central government and EU institutions to look at Britain's position in European environmental policy making. Part III looks at the various institutions – environmental pressure groups, local authorities, environmental agencies and industrial and business lobbies – involved in UK environmental policy to see how they have responded to Europeanisation. Part IV then takes separate policy fields – nature conservation, land use planning and landscape protection, water policy, waste management and industrial pollution – to examine how practices, procedures and structures have been changed. Finally, in Part V, a concluding chapter seeks to draw together the lessons to be learnt for the future strategy of the British environmental sector in Europe.

Part II

BRITAIN'S POSITION IN EUROPEAN ENVIRONMENTAL POLICY MAKING

Part II concentrates on the key axis between national government and European institutions. The chapters focus on broad governmental strategies in response to integration in the environmental sector. The major tasks of this part are threefold: to establish the determining factors shaping British diplomacy; to explain how and why British–EU environmental relations have changed over time; and to examine the provenance of Britain's poor environmental reputation (outlined in Chapter 1) and how British governments have tried to counter it.

In dealing with these tasks, the three chapters in Part II assess Britain's position from different vantage points. The first two chapters by Robin Sharp and Nick Hanley analyse Britain's response to integration from either end of the European telescope, Sharp from a Whitehall perspective and Hanley from that of the European Commission. Both authors draw on their personal experiences as insiders in European environmental policy making to illustrate their arguments.

As a recently retired civil servant, with more than twenty years' experience in the upper echelons of the Department of the Environment, Robin Sharp is well placed to recount central government's evolving reaction to Europeanisation (see Chapter 2). He illustrates the broad changes in ministerial and civil service predispositions by reference to three Directives (Habitats, Urban Waste Water and Integrated Pollution Control). Each of these Directives provides evidence of a shifting British response to the European policy dimension. Essentially, Sharp argues that Britain has moved from an initially complacent attitude to a more proactive role in trying to shape the European environmental agenda. Despite this more positive stance, Sharp concludes that Britain is still a middle-ranking member state which mostly adapts rather than initiates in the environmental arena.

In Chapter 3, Nicholas Hanley, having worked in the European Commission for over a decade (after having trained and worked as a planner in Britain), reflects on the role of governmental and non-governmental actors from Britain in the different stages of the European policy process – agenda setting, formal discussion, decision making and implementation. He concurs with Sharp that Britain has increasingly adapted to Europeanisation to play a more active role, notably in the field of implementation politics. Indicative of this increasing adaptation is transnational coalition building by all British policy actors. A significant consequence, Hanley concludes, is that it has become increasingly difficult to identify purely national interests and policy actors in the European policy arena.

Of course, other member states are also experiencing the process of European integration, even if the consequences are nationally differentiated. To highlight what is distinctive about the British experience, there is a need to place it in a comparative context. Henry Buller's chapter is an Anglo-French comparison of the domestic Europeanisation of environmental policy (Chapter 4). Britain and France are commonly identified as middle-ranking countries in the development of EU environmental policy, and more recently the two governments have adopted similar concerns about implementation, deregulation and subsidiarity (see Chapter 1). Even so, the national ramifications have been divergent.

Buller examines two important aspects of integration: the broad adaptation strategies of member states and the domestic administrative consequences of Europeanisation. He contrasts the British and French experiences on both counts. In the case of the former aspect, whereas France's approach has been one of 'internalising' EU environmental policy, Britain's has been to 'externalise' it. The differences between existing British practice and European initiatives have been critically emphasised in an effort to maintain a separation between the domestic and European spheres. In the case of domestic administration, Buller argues that integration has encouraged a process of regionalisation of environmental administration in both countries. However, because of the different governmental traditions, Europeanisation in Britain is viewed (and criticised) as a centralising process, whereas in France it is often associated with decentralisation and with positive connotations.

2

RESPONDING TO EUROPEANISATION

A governmental perspective

Robin Sharp

Without attempting to be comprehensive, this chapter seeks to highlight the evolution of the UK Government's approach to environmental policy issues in the European Community, relating it to changing attitudes on the part of Ministers and officials in the Department of the Environment (DoE), the lead department concerned. It is an insider's impression and is based partly on direct experience, especially during the period 1991–5 when I was the Director supervising the DoE Divisions responsible for, inter alia, the Habitats Directive and the co-ordination of European Union environmental work.[1] The chapter reviews the experience of negotiating some specific instruments during this period, including the Habitats Directive, as well as the Government's response to the Fifth Environmental Action Programme. It also examines efforts to improve the European professionalism of the DoE. The chapter concludes that under strong ministerial leadership the UK is now seeking a positive role in European environmental policy making and is deploying its resources more effectively to that end than was the case previously.

THE 1970s AND 1980s

It is clear that we have come a long way from 1972 when the UK signed the Treaty of Accession. As Special Assistant to the then Geoffrey Rippon in the Commons and Lord Jellicoe in the Lords, I had the dubious privilege of listening to virtually every word spoken in Parliament during the prolonged passage of the European Communities Act 1972. I cannot recall any mention of the environment, not just in the Bill but in all the hundreds of thousands of words that were thrown across both Chambers in order to debate the biggest constitutional change affecting Britain since the 1707 Act of Union.

Perhaps we should not be surprised at this since the Treaty of Rome made no mention of the environment. It was above all an economic treaty made for profoundly political purposes. Yet Community legislation with clear environmental consequences was being enacted even in these early days for the UK. It was authorised by the Treaty provisions designed to remove technical barriers to trade. This, of course, is an economic objective, aimed at breaking down the use of higher standards or complex procedures to protect domestic manufacturers. Such standards or procedures are normally justified by safety or environmental considerations.

Both types of consideration affected the standards for motor vehicles and, by chance, I found myself working on vehicle safety standards on return to the DoE after the European Communities Act had reached the statute book. A colleague was responsible for vehicle pollution and he, also, spent a great deal of his time in Brussels negotiating new Directives to allow motor vehicles to be traded freely throughout the nine member states. However, our primary motivation was to raise safety standards in order to save lives and reduce injuries. We did not start from the premise that UK standards were perfectly satisfactory and thus that our objectives should be to persuade the other eight to rally to our position. A common aim was to improve road safety, and to achieve this we had to push and pull the manufacturers and their Industry Department minders to incorporate cost-effective safety improvements into their new models.

These personal experiences demonstrate not only that environmental policy in the European Union can be and often is authorised by economic provisions of the Treaties, but also show that shared assumptions about key objectives and peer pressure can be a powerful force for raising standards generally. This argument in support of common action is not one often heard during a phase when influential elements within the UK are having doubts about the scope of Community legislation. Doubtless its importance varies from time to time, depending on attitudes of key countries and the European Commission, but it should never be underestimated as a significant motivation for framing environmental legislation at the European Union level.

Although much formative environmental legislation was adopted in the 1970s and 1980s by the Community it is fair to say that the process of making it was not accorded great priority by the DoE (see Chapter 1). Generally the substance of Directives was negotiated in Commission and Council Working Groups by officials at Grade 7 (then Principal) or Grade 5 (Assistant Secretary) or their scientific equivalents. More senior grades rarely attended such meetings or visited Brussels. Nor, apart from those with specifically European responsibilities, did they accompany Ministers to the Council where legislation was formally adopted. It was more important to stay at home and focus on Ministers' domestic agenda. On the whole few Secretaries of State attended meetings of the Council: instead they sent their Ministers of State.

This climate of general inattention by the Department's most senior Ministers – they had the abolition of the Greater London Council and taking on local government expenditure as more serious issues to tackle – undoubtedly influenced the way in which officials negotiated in Brussels in the early 1980s. Moreover, there was a period of distinct Euro-scepticism before that term entered the political debate. In such an atmosphere there were two likely outcomes and both had their place. Where Ministers' views were sought the instructions to negotiators tended to be restrictive. In other cases there was a judgement that they would not want to be bothered with matters that seemed at the time to be predominantly technical. A third factor was also at work: a sincerely held belief that member states had considerable freedom in implementing directives into their own legislative and administrative structure. When the Commission and the European Court of Justice took a different view this led to a great deal of angst in Whitehall generally, including the DoE. It was not until the beginning of the 1990s that a specific co-ordinating division for European environmental work was established, reflecting not only the increased level of Community activity in this sphere, but also the seriousness of the Department's approach to it.

The legacy of the 1970s and 1980s is still with us but to a large extent we have moved into a new era. The impact of Mrs Thatcher's espousal of the global environmental agenda and Chris Patten's 1990 White Paper followed by John Gummer's arrival as Secretary of State in 1993 brought about a step change, although not a revolution. The tangible evidence for this and the outlook for the future are considered below, but first it is helpful to move from the assertion of generalities to an examination of the negotiation of

Table 2.1 UK Secretaries of State for the Environment

	Secretary	*Party*
1970	Peter Walker	Conservative
1972	Geoffrey Rippon	Conservative
1974	Tony Crosland	Labour
1976	Peter Shore	Labour
1979	Michael Heseltine	Conservative
1983	Tom King	Conservative
1983	Patrick Jenkin	Conservative
1985	Kenneth Baker	Conservative
1986	Nicholas Ridley	Conservative
1989	Chris Patten	Conservative
1990	Michael Heseltine	Conservative
1992	Michael Howard	Conservative
1993	John Gummer	Conservative
1997	John Prescott	Labour

specific instruments to see what light they throw on the UK's underlying attitude to and competence in negotiating law in Europe.

The instruments selected here are the Habitats and Species Directive of 1992, which illustrates well changes in the UK's attitude; the Urban Waste Water Treatment Directive, because of its transmogrification from a good news story into a substantive political concern which raised threats to the Conservative Government's electoral position; and the Integrated Pollution Prevention and Control Directive, as an example of a UK negotiating success.

THE HABITATS AND SPECIES DIRECTIVE

The Directive on the Conservation of Natural Habitats and of Wild Fauna and Flora, 92/43/EEC, is of great interest, not merely as an instrument of nature conservation, but as an example of European Union law making and evolving UK Government attitudes. Because birds attract a great deal of popular interest and because many of them migrate, it was not surprising that in 1979 the Community took its first step in nature conservation by adopting the Birds Directive, requiring member states to protect them along with the breeding and feeding sites of threatened and migratory species. In the UK this was implemented in somewhat haphazard fashion through the Wildlife and Countryside Act 1981 (see Chapter 12).

It was not until 1988 that, stimulated in part by British conservation groups (most notably the RSPB and the Mammal Society), the Commission finally delivered to the Council of Ministers a draft directive designed to be an EC implementation of the Bern Convention, that is, to do for plants and other animals what the Birds Directive had done for birds, but also including a category of endangered semi-natural habitats irrespective of whether they contained threatened species. It was this novel feature of the proposals which led the DoE always to refer to it as the 'Habitats Directive' and, as a consequence, to focus on the site provisions. The Commission's document lacked eight of the eleven annexes referred to in the substantive articles, making assessment of its impact almost impossible. In October 1989 the French Presidency tabled a new compromise text, removing the power of the Commission to designate sites without member state agreement. In March 1990 the Commission supplied the annexes missing from its original proposal. *Inter alia*, the lists of species contained a substantial number present in the UK which were not at the time protected or for which management plans would be required. Following proposals by the European Parliament for fifty-three amendments to the text (largely reflecting the views of conservation groups), the Commission submitted a revised text to the Council in February 1991, accepting twenty-nine in whole or in part. Negotiations then gathered pace. A restructured proposal was tabled by the Luxembourg Presidency in May 1991 and debated by Ministers in the

36

subsequent June Council. At that point, the main substantive issue became the question of special financial provisions to assist implementation in nature rich but economically poor member states (principally Spain) and final political agreement was only reached at the December 1991 Council after two more rounds. Formal notification of the text of the Directive did not occur until 5 June 1992 which then became the starting gate for a series of deadlines for the Commission and member states culminating on 5 June 2004.

As the House of Lords Select Committee on the European Communities report (1989) on the proposal pointed out, the initial reaction of UK Ministers to the Directive was unenthusiastic. The Committee regretted 'the somewhat parochial attitudes' which had characterised the Government's evaluation of the initiative from the outset and referred to a speech by the Minister of State (paragraph 75). Among other somewhat chauvinistic remarks in his speech, Lord Caithness had said:

> We really must keep control of the future of our own countryside and of our wildlife in our own hands . . . by what is no doubt an administrative oversight, however, [the draft Directive] fails to propose the addition of the Brussels bureaucrat to the official pest list! . . . I have to say we do not like the draft Directive as it stands.
>
> (1988)

The Committee thought Government should instead be turning the UK's expertise in nature conservation to positive effect at the European level and believed some other member states were looking to us to give a lead.

The Committee identified from the evidence taken from others 'a growing sense of a "European" heritage of flora and fauna, the protection of which should be a collective Community responsibility' – a feature of the Commission's proposal was the establishment of a coherent European ecological network of sites of Community importance to be known as Natura 2000. However, the Government appeared to see 'its nature conservation obligations in almost wholly national terms, albeit tempered by a relatively good record in meeting commitments under various international conventions'. Doubtless the Department's official perspective was heavily influenced by the views of the then Secretary of State, the late Nicholas Ridley, who was not only Euro-sceptical but also a strong adherent of the traditional countryman's approach to managing wildlife, as opposed to the bureaucratic regulatory approach, whether emerging from Brussels or Whitehall. Nevertheless, it was not unfair of the Committee to detect from departmental officials a view that our protective systems were well developed and among the best, so why have all the bother of superimposing a set of European requirements whose practical benefit would largely be to improve the prospects for nature conservation in more backward member states? Since plants and habitats are not migratory and other animals, apart from marine fish, less migratory than birds, the conservation lobby also felt that the UK should be pro-active for the wider benefits that would bring, rather than that it would make a real difference here.

Another factor influencing officials' approach to the draft Directive was their experience in, and ministerial attitudes towards, the implementation of the Birds Directive. Among other things the latter required specific licensing arrangements for the killing of so-called pest species such as crows and sparrows but this was not provided for in the implementing Act of 1981. As the Commission began to press its case against the UK to remedy the deficiency, Conservative Ministers and their country friends found it massively irritating that, as they saw it, Brussels bureaucrats were trying to interfere in the minutiae of traditional English countryside sport and pest control. At the same time, very slow progress had been made in designating the Special Protection Areas (SPAs) required under the Directive, partly because the Nature Conservancy Council was preoccupied with reassessing existing Sites of Special Scientific Interest (SSSIs) and spreading the network as fast as it could and was reluctant to provide the scientific underpinning needed, and partly because the Department took the view that *de facto* protection for potential SPAs was provided by the SSSI and development control systems, making the UK proof against any European Court of Justice proceedings about the absence of formal designations. Against this background the watchword was to keep a low profile and proceed with caution. All this tempered enthusiasm for the fresh wave of provisions heralded by the new Directive.

By the time I became directly involved in May 1991, a number of changes were in the wind. New Ministers such as Michael Heseltine and David Trippier were taking a much more forward stance on environmental issues generally than their 1988 predecessors. *This Common Inheritance* had been published by the Government in 1990 and the DoE at any rate was looking for positive outcomes in European negotiations. Thus in the spring of 1991 we find Michael Heseltine telling an all-party Parliamentary back-bench committee, in reference to the Habitats Directive, that 'the UK is fully committed and indeed enthusiastic about its concepts' while calling for realism and avoidance of anything resembling the irksome pest bird provisions of the Birds Directive (text of speech, Spring 1991).

It was in this spirit that the UK at ministerial and official level negotiated in the Council during 1991. The main UK worry was to deal with the draconian consequences of the European Court of Justice judgment in the *Leybucht* case (57/89) which emerged in 1991 and held that, under the Birds Directive, damage – for example through development – to a designated site could only be justified on grounds of human health and safety and not merely on social or economic grounds. This judgment had confirmed the worst fears of economic Departments and the European lawyers in Whitehall that the Directive would remove discretion from member states in cases where development proposals affected Special Protection Areas.

The Department accordingly had to frame amendments to the Habitats and Species Directive, not only to ensure that social and economic considerations could be taken into account but also to apply this regime to the Birds Directive. The aim was to modify the effect of the *Leybucht* judgment, though without opening up a loophole which would allow the nature conservation interest to be overruled willy-nilly. It was considered that without some flexibility it would be very difficult to secure agreement to the designation of many sites. A suitable formulation was brokered through the Whitehall official machinery of EQL (the legal group) and EQO (the policy group) before being tried on the Presidency. Although aspects of British drafting turned out to be too sophisticated to be followed by some member states and the Commission was anxious to limit any changes in reaction to *Leybucht*, the initiative was successful. At an operational level it was also important in ironing out differences between the protective regime in the Birds Directive and that proposed for Special Areas of Conservation (SACs) in the new Directive. Since many sites would qualify for designation under both instruments it would have been intolerable to have had two differing regimes governing them, but the Commission seemed to have given little thought to this.

The last significant hurdle to the adoption of the Directive was financial. Since the Commission had made no attempt to identify the likely extent of sites to be designated as SACs or of any management or restoration plans required, there was no information available about the order of costs of implementation. No one could contest the Spanish assertion that their country was disproportionately rich in endemic species and threatened or endangered habitats. It therefore seemed probable, if unproven, that significant costs might fall on them and they developed the line that if the rest of the Union wanted them, as one of the poorer member states, to preserve their natural heritage, there should be a contribution to the costs, or 'Community co-financing' as it was called. Notwithstanding the existence of a separate modest financial regulation for nature conservation (under LIFE[2]), the Spanish held out for an express reference to co-financing in the Directive.

While such a general reference might have seemed reasonable to the DoE, who accepted that there was a case for modest co-financing as long as the Spanish accepted the primary responsibility for looking after their own habitats, it was anathema to the Treasury as the guardians of UK financial orthodoxy in European Union business. They much disliked any breach of the 'polluter pays' principle, which in this context they took to mean that each member state should shoulder the domestic effect of agreed legislation. In general, the Treasury also opposed any financial references in Community texts that were not under the direct control of ECOFIN, the

Finance Council, and might be used to put pressure on the overall Community budget. I can well remember negotiating on this in a parallel working group of the Council in June 1991 and discovering that my every remark was immediately phoned down the line by a colleague from UKRep[3] to a Treasury official in London, who in turn sent back instructions as to what I could and could not say. In the end there was no agreement until December; the Spanish got more than they deserved; but Britain did extract from them, and get into the text, the important concession that they must prevent damage to priority sites, even if they could not be required to apply a proactive management regime in them until Community funding was available.

Altogether this represented a very satisfactory package for the Government and the DoE in particular, but there was almost a slip at the last moment when a technical working group tidying up the annexes proposed the addition of some obscure plant species which did not occur in the UK but which had not been cleared by the Scottish Office. As so often in such cases the written briefing was too delphic for the negotiators to know whether it mattered, and after working level contacts by telephone between Brussels and Edinburgh had drawn a blank I became involved in London. I discovered that the only responsible official left in the Scottish Office Environment Department was the Permanent Secretary and I spoke to him. He in turn found out that the ecological expert who had raised the query had taken an afternoon's leave and could not be reached. I had to point out that the British Minister at the Council table could not possibly hold up political agreement on a measure Britain was now very keen to see enacted on such a technicality and so two people, neither of whom had the faintest idea what the problem was, took a decision to accept inclusion. So is European legislation made. This bizarre episode does illustrate another point: although the UK is a unitary state the territorial Departments can sometimes take a semi-detached view of Community proposals and nowhere more so than in the field of nature conservation.

The reasons why territorial Departments may from time to time sit loose to European regulations and the implementation of the provisions ultimately adopted are complex. In the case of Scotland, pressures from Opposition parties for constitutional independence or devolution were countered by a response from Conservative Ministers that a tartan shading should be applied to the Government's own policies where feasible. This encouraged their officials either to be cautious in giving their agreement to proposals, particularly those affecting land management, which seemed likely to cause domestic controversy, or occasionally to entertain the idea that there could be special provisions for their area. In addition, shortage of resources and the impracticability of having too large a negotiating team in Brussels make the presence of a territorial official a rare event and it is always easier to say 'no' when you are not sitting round the table.

In general, the priority of Scottish and Welsh Departments is to secure successes for their native patch. European legislation (as opposed to exploitation of the EC Structural Funds[4]) does not score very highly in this index. There is therefore an inevitable tendency for territorial colleagues not to follow all the details of a long negotiation or to come up at the last moment with points which the lead negotiators would have preferred to deal with at an earlier stage. On the other side of the coin the lead negotiators may, as a result of time pressures, not always be diligent in reporting to territorial colleagues all the twists and turns of negotiations which might be of concern. Two-way communication requires both time and a readiness to share information. The former is in short supply, and increasingly so as the Departments are progressively 'downsized' to meet other Government objectives.

Unlike the Birds Directive, the Habitats and Species Directive lays down a very specific timetable for the various stages of implementation, starting with the enactment of domestic legislation by 5 June 1994. Given the extent of existing legislation implementing the Bern Convention on which the Directive is based, this target looked relatively easy to meet, but it was not to prove so. Thinking in the Department was that primary legislation was desirable to allow for debate on the sensitive issues involved in the restrictions which site designation would impose on private landowners, even though the sites in question were already SSSIs. The Government's business managers were unsympathetic to this view, seeing the measure as potentially troublesome in the Lords. They preferred the use of secondary legislation under section 2 of the European Communities Act which, contrary to the normal rules, allows such secondary legislation to amend or override primary legislation if needed to give effect to a Community obligation. In this they were supported by the EQL legal network in Whitehall whose argument was that developing judgments of the European Court of Justice required very precise and comprehensive implementation of the text of the Directive, which could only be achieved by secondary legislation not open to amendment in its parliamentary consideration. This was a far cry from the heady days of 1972 (or the 1981 Act) when the Government firmly believed that, unlike Regulations taking direct effect, Directives afforded member states plenty of room for flexibility when it came to transposing them into national law.

Having had its sights trained firmly on to implementation through regulations under the European Communities Act, the Department then discovered that it was necessary not only to adjust the site safeguard provisions of the existing domestic legislation but also to confer duties to act in accordance with the Directive on all relevant Secretaries of State and Government Agencies and to modify a huge raft of powers of public bodies which might potentially come into conflict with it. Moreover, there was no effective protective nature conservation regime in the marine area, nor did the Marine Transport or Fisheries bureaucracies readily accept that there should be, whatever the Directive might say. It also became clear that more had to be done on the species aspects than had been thought when the focus of the debate was echoing the Department's shorthand of referring to the instrument as the 'Habitats Directive'.

In spite of all this we were surprised to discover that the parliamentary draughtsman, to whom the task of framing regulations under the European Communities Act is given, produced of the order of 100 pages of text to give effect to the Directive and even that did not cover all the requirements. That the Regulations (1994 Statutory Instrument No. 2716) were laid before Parliament, if not actually approved, only a few days after the 5 June deadline represented a heroic effort on the part of key officials, again almost thwarted at the last fence by a territorial intervention, this time from the then Welsh Secretary. In achieving this result and moving forward with the process of forwarding lists of selected sites to the Commission, the UK has assumed a leading position among member states, making amends for its leisurely and incomplete implementation of the Birds Directive during the 1980s.

What lessons about UK policy can be learned from this unfinished saga? First, there was a distinct evolution of the UK's overall attitude as the negotiations progressed. Part of this can be fairly attributed to improvements and clarifications which were made as a result of negotiation: the original text was seen as mechanistic and impracticable, as well as giving excessive powers to the Commission. Nevertheless, the arrival of new Ministers with more positive attitudes to Europe and to conservation legislation also made a difference. What could have been convenient sticking or delaying points were treated as obstacles to be overcome and a sense of momentum was engendered at both the political and technical levels. The handling of the effect of the *Leybucht* judgment demonstrates the interest of other key Whitehall Departments in conservation legislation affecting their policies and practices. The Treasury's involvement in the Spanish bid for express reference to Community financing reflects the consequences of Mrs Thatcher's policy of reducing the net UK budget contribution to the minimum possible.

A major irony of the overall outcome is that, whereas at the outset British conservationists and the Department were taking the view that the main benefit of the Directive would be to bring some of the laggard Members up to scratch, the actual effect on UK legislation has turned out to be substantial. In practice the SSSI protective regime embodied in the 1981 Act and its predecessors is essentially discretionary for Government. In the last analysis, development or agricultural damage can destroy the nature conservation interest of a special site, even one of international importance, if the Government considers that that is where the balance of advantage lies. There is no absolute protection or indeed presumption against development in the primary legislation. While the Government's guidance to planning authorities was largely in line with the Directive, it could always be changed or re-interpreted. In contrast, the Directive contains a rigorous protective regime for the sites concerned, greatly increasing the obstacles to be over- come if damage is to be allowed, and puts duties on Ministers and others to further its objectives, thus exposing them to action in the UK courts as well as the European Court of Justice if they fail to do so. Such action has already been pursued by the RSPB under the Birds Directive (Plate 12.1).

It is to the credit of the recent Ministers that they did not flinch from this but have spoken positively and openly of the value of the European natural heritage and the part the Directive can play in halting and reversing its erosion. In launching the UK consultation on the potential list of sites to be forwarded to the Commission, John Gummer, the then Secretary of State, described it as 'one of the most important milestones for conservation in Europe' (DoE 1995d). A glossy explanatory booklet was produced to accompany the consultation papers (DoE 1995b). Perhaps the day is not too far off when the selected sites themselves will have display boards bearing the geese and stars of the Natura 2000 logo and an explanation of the special contribution they are making to European and global biodiversity.

THE URBAN WASTE WATER TREATMENT DIRECTIVE

Published by the Commission in 1989 and agreed by the Council of Minis- ters in 1991, the Urban Waste Water Treatment Directive (91/271/EEC) has had major effects on UK policy and practice, some intended and some not originally foreseen (see Chapter 14). Its main objective is to protect the environment from the adverse effects of waste water discharge from homes, industrial plants and surface run-off. The main adverse effects are deoxygenation leading to dead rivers and eutrophication of fresh and coastal waters, although there are also public health and aesthetic implications from the disposal of untreated sewage.

In order to raise standards, treatment is necessary and the principal thrust of the Directive is to require secondary treatment (involving some biological process) to discharges to fresh and estuarine waters from urban areas above 2,000 population by 2005 at the latest. There are more stringent demands for sensitive areas and relaxations for coastal and estuarial areas where, in many countries, no treatment arrangements existed when the proposal was made. At that time there were very wide variations both between and within member states. While Portugal discharged 80 per cent of its sewage untreated, the UK was treating 80 per cent of its inland sewage to secondary standard but most of its sewage from coastal towns received only a pre- liminary form of treatment. The UK was responsible for almost all dumping of sewage sludge at sea by EU countries.

Plate 2.1 Industrial outflow into the River Ouse, Humberside. During the 1980s
Britain came under mounting international pressure to clean up its indus-
trial and urban discharges into the North Sea.
Source: Greenpeace/Greig

Although the treatment of sewage and other waste water is not the common currency of conversation in pubs and clubs, the Directive has a clear political pedigree, flowing as it did from a ministerial meeting on Community water policy held in Frankfurt in 1988. Nor can it be doubted that the success of the Greens in the European Parliament elections in 1989 and the criticism of the UK that was to be expected at the North Sea Conference at the Hague in 1990 led Chris Patten, then Secretary of State for the Environment, and David Trippier, his Minister of State, to announce decisive shifts in UK policy just before the Conference.

First came the undertaking that all substantial discharges of sewage into the sea would generally be subject to treatment, and then the announcement by the Ministry of Agriculture, Fisheries and Food that the dumping of sewage sludge at sea would end by 1998. David Trippier told the Commons Environment Select Committee that:

> the use of long sea outfalls has never been accepted enthusiastically by the public and neither has it received universal acclaim by the world scientific community. Doubts have persisted about reliance on the sea to purify and to render harmless sewage discharges, and there has been concern about the potential long-term build-up of pollutants in the sea.
> (House of Commons Select Committee on the Environment 1990:
> col. 1201)

Thus it was clear that Ministers were driving the main perceived changes to UK policy and practice which would be needed to comply with the Directive. This was not a proposal devised in a corner by technocrats who failed to explain its significance. It also received thorough scrutiny by the House of Lords Select Committee on the European Communities (1991) who criticised the reliance on uniform limit values rather than environmental quality objectives and a failure to assess the environmental consequences of disposing of sewage sludge through landfill and incineration. Nevertheless, taking account of the fact that some UK amendments were accepted and important political objectives were secured, the Government and the Department appeared to have good cause to be satisfied when the Directive was agreed in March 1991.

At that time the best estimate of costs was obtained from consultants who advised the Department that improvements in coastal and estuarial disposal facilities would cost of the order of £2 billion in addition to the £1.4 billion programme for long sea outfalls already accepted to comply with the Bathing Water Directive. In the context of an investment programme of some £24 billion over ten years incorporated in the water industry privatisation prospectus, this addition did not appear to pose a problem. Nevertheless, costs of implementation became an issue when, in July 1993, the Office of Water

Services (OFWAT) produced an estimate of £10 billion for the investment required in England and Wales only, with dramatic implications for water charges. After further consideration of the methods of implementing the Directive, OFWAT's estimate was reduced to £6 billion. The regulator's message was that environmental improvements were being bought at too high a price. MPs with constituencies in the South West, where the greatest effects would be felt, were directing their complaints to the Prime Minister, and the Department began to put out feelers to the Commission to obtain amendments to the Directive which would allow more time for the investment to be made (Ward, Buller and Lowe 1995). Although other member states are also belatedly waking up to the fact that implementation costs are much higher than they had originally imagined (see Table 9.1), the Commission has so far remained unmoved.

During negotiations the Department appears to have relied on the consultant's figure of £2 billion for coastal improvements and accepted the view that inland secondary treatment would pass muster or be covered in the overall £24 billion agreed investment programme. The political imperative behind the original proposal in the national and the international context and the belief that many of the requirements did not go beyond those allowed for under privatisation, plus the adoption of UK amendments designed to secure other cost reductions, may have led to a too superficial assessment of the technical detail of the limit values. The report of the House of Lords contains a footnote reporting a Commission estimate for compliance by the Community as a whole of £28 billion and a German Government one of £140 billion, both unconfirmed (1991: 51). One moral of this story is the need for the Commission to be obliged to produce proper costings for its proposals, as is required for Public Bills in the UK (see Chapter 9). Doubtless it has made UK negotiators even more careful to establish implementation costs beforehand and to use them as a negotiating tool. There must be considerable sympathy for the House of Lords' view that, with such large expenditure at stake, greater cost-effectiveness could have been achieved by using the environmental quality objectives approach rather than the moderated form of uniform emission limits in the Directive. How much a reduction in deoxygenation and eutrophication might be worth, as against other environmental gains which similar expenditure could secure, is another question which could have been more thoroughly debated.

Heightened concern about the likely costs of the Urban Waste Water Treatment Directive and the interaction between it and other EU water legislation, including the draft Ecological Quality of Water Directive, have led to an enhanced level of dialogue among EU member states and Commission officials responsible for water policy. This has been actively promoted by the Head of the Water and Land Directorate in DoE, who has forged contacts with opposite numbers in most member states and the Commission Director to the extent that there are now regular meetings of

Water Directors under the Commission's chairmanship. It is to this group, with the UK as one of the key pressure points, that we can look for the origin of the Commission communication reviewing EU water policy and proposing a Water Framework Directive which would repeal some existing measures and consolidate others, including the draft Ecological Quality proposals (CEC 1996e). Even though the Commission are not proposing to amend or integrate the Urban Waste Water Treatment, Drinking Water and Bathing Water Directives, the review and the Framework proposal must be seen as an important step in the right direction and much to be welcomed. The communication can be seen as a success for the UK because it reflects a strategic approach and incorporates the principles of the Fifth Environmental Action Programme discussed below.

THE INTEGRATED POLLUTION PREVENTION AND CONTROL DIRECTIVE

The Integrated Pollution Prevention and Control Directive (96/61/EC) provides a good example of a measure which the UK Government has been anxious to promote from the word 'go'. It is principally directed at large, potentially heavily polluting industrial plant affecting more than one of the media – air, water and land – and is designed to ensure, through a site-specific permit system, that any unavoidable pollution is directed to the part of the environment best able to receive and deal with it. Such a system was implemented in the UK through the Environmental Protection Act 1990 and is operated by Her Majesty's Inspectorate of Pollution (HMIP), (now part of the Environment Agency).

The integrated pollution control (IPC) system in Britain reflects the British approach to tailoring controls to specific sites where possible, taking account of the ability of the media to absorb the pollutants in question, and employing the best available technology not involving excessive costs (see Chapter 15). This achieves the maximum environmental benefit for each pound spent, in contrast to systems where limits are set nationally and for each of the media separately. In promoting this philosophy in Europe the UK was therefore seeking to advance its rational attitude to environmental regulation, as a precedent for future legislation. It was also keen to secure a level playing field for British industry by ensuring that large polluting plants throughout the Union were treated in a similar way (see Chapter 9).

When the Commission's intention to bring forward a proposal was confirmed, the Department of the Environment offered one of its administrative officials as a detached national expert to be seconded to DGXI to work on the drafting of the Directive and this offer was accepted. The resulting instrument, as transmitted by the Commission to the Council in September 1993 (CEC 1993b), was regarded as highly satisfactory by the Government, as far as the main articles were concerned, since not surprisingly it was very close to the British framework. There were, though, problems for others. The Commission's explanatory memorandum revealed that a number of countries, including Germany and the Netherlands, had a strongly embedded position of regulating emissions to each environmental medium separately, while Spain left control of air and waste to regional authorities, and water was regulated by a national agency (CEC 1993c: 6–9). The coverage of the Directive posed some problems for the UK too, as it included sectors such as farming and the food and drink industry not currently covered by IPC here. This reflected the scope of IPC in some other countries, such as France, who had been operating it for some time.

After some fairly unstable negotiations under the German Presidency, during which the Department was unable to furnish a coherent text to the UK Parliament, the French Presidency took matters in hand and a delicately balanced common position was achieved at the June 1995 Council. Although the two sectors of farming, and food and drink, which the UK would have preferred to have been dropped, were still included, various ameliorations were secured to lessen the burden on them.

Appearing before the House of Commons European Standing Committee, James Clappison, a Junior Environment Minister, was emboldened to say:

> June's decision gives the United Kingdom a complete victory on those major points of principle. The draft directive sets out a framework for European policy on the prevention of pollution from large industrial plants under which a site-specific approach will be used, which takes account of the environment as a whole, and adequate prominence will be given to the consideration of cost-effectiveness and economic feasibility. That framework is fully consonant with UK environmental policy and is very much to be welcomed.
>
> (House of Commons European Standing Committee A,
> 13 December 1995: col. 3)

It is interesting to note that in the briefing sent to UK MEPs in October 1995 the Department pointed out a number of minor amendments which would meet UK objectives, coupled with a strong message that anything too radical in the way of amendments to the articles might rock the boat and risk losing the whole Directive (DoE 1995c). The final Directive which was agreed in October 1996 represented a significant step towards an integrated rational approach to regulation, compared with the rather fragmented hit-and-miss style which had characterised a fair amount of the Commission's proposals in earlier years. Although modelled on the UK approach, the Directive is somewhat broader in its requirements. The case for including intensive farming in IPPC proved as intellectually defensible as the long-term British policy of exempting the industry from planning controls was ultimately indefensible. Inevitably some trading on detail is necessary if a major policy gain is to be achieved.

THE EC FIFTH ENVIRONMENTAL ACTION PROGRAMME

It is fair to say that the UK did not attach much weight to the first four environmental action programme documents produced by the Commission. They were seen as little more than somewhat imprecise shopping lists of proposals for future legislation whose scope and timing were largely shrouded in mystery. It was not easy to discern a theme for the Commission's programme up to the end of the 1980s, though the quantity of legislation enacted rose rapidly towards the end of the decade (see Figure 1.1). As public support for environmental protection measures grew, the impression was that the Commission looked around for gaps that could be quickly and easily filled. Moreover, it was not until 1995 that a comprehensive European state of the environment report was published, which could offer scientific underpinning to a more strategic and prioritised approach (Stanners and Bordeau 1995).

However the Fifth Environmental Action Programme, published by the Commission and endorsed by the Council in 1992 (CEC 1992b), has been taken fairly seriously by the UK, reflecting both its more substantive content and developments in the UK approach to EU environmental policy. The very title of the Programme, Towards Sustainability, is drawn from the new objective in Article 2 of the Maastricht Treaty on the 'promotion of sustainable and non-inflationary growth respecting the environment'. Not by chance does it complement the Rio Earth Summit declaration on environment and development and the detailed programme for sustainable development set out in Agenda 21.

The UK could not do other than welcome the principles set out in the Fifth Programme, including shared responsibility among relevant actors, the precautionary approach, 'polluter pays' and the full integration of environmental considerations into all aspects of policy making. After all, the UK was one of the first governments after the Rio Earth Summit to produce substantive policy documents explaining how it proposed to carry forward the major commitments it had made there. It could not therefore cavil at the Commission seeking a similar broad agenda, moving away from a mere list of environmental regulations to an attempt to engage sectors such as agriculture, industry, energy and transport to make their activities more environmentally sustainable.

Tangible evidence for the more serious UK approach can be found in the DoE's 1994 document *Towards Sustainability: Government Action in the UK*, as a contribution to the Commission's mid-term review of the Programme (DoE 1994). This reports in schematic, line-by-line format, on UK Government actions in the areas covered by the programme and contains a brief annex on the local government responses. The task of producing the document was taken up less than enthusiastically within the Department, not because of embarrassment at inactivity in one sector or another, but through sheer weariness at the plethora of reports being called for following the Rio Earth Summit and scepticism as to who the readers might be. In fact it was pressure on Ministers from the environmental lobby which caused the work to be done, but with hindsight it appears to be a small but useful step in giving the UK Government more confidence about the Fifth Programme and its relevance as a strategic tool at the European level.

Furthermore, the report focused the Government's attention on elements in the Programme which it wanted to see developed. In the introductory paragraph the Government emphasises the strengthening of the enforcement network IMPEL, linkages with the European Environment Agency to strengthen practical implementation, the integration of environmental policy into the Common Agricultural Policy and the Common Fisheries Policy, the Trans-European Networks and the Structural and Cohesion Funds. More detailed Government views were sent to the Commission in an unpublished letter before the Commission forwarded a draft decision on the review (CEC 1996a) for adoption jointly by the Council and the European Parliament. This is a novel procedure which may lead to some interesting negotiations (see Chapter 6). The UK Government, while considering its position on the detailed substance of the draft decision, welcomed the strategic and integrating approach which had emerged even more strongly.

The 1994 UK report also mentions three new EU environmental fora which have been created under the Fifth Action Programme. The first is the so-called Environmental Policy Review Group consisting of the Director-General of DGXI and the equivalent senior civil servants of all the member states. The Group meets regularly and, although only consultative, provides a good opportunity for informal policy discussions, for example, on the review of the Programme, work in hand in DGXI and work on sustainable development involving other sectors. Although the Environmental Policy Review Group has established useful working groups on agri-environmental and transport issues and is regarded positively by the UK, it is not yet seen as having delivered its full potential. The view from London is that more could be done by way of frank discussion of the success or otherwise of implementing existing measures at the member state level and by close encounters with other Commission Directorates responsible for the sectoral policies affecting sustainable development. At the working level the UK is a keen promoter of enforcement know-how through the new IMPEL network (see p. 99) which builds on a UK initiative. The third new body is the Consultative Forum on the Environment on which sit environmental groups, business and local government representatives. So far the Government has had little feedback from the Forum and is somewhat agnostic about its usefulness.

Nevertheless the Fifth Action Programme and its review process, coupled with the new fora recently created, all promote a more strategic approach to the environmental agenda which is most welcome to the UK Government. It remains to be seen if dialogue with agriculture, transport and energy can overcome the formidable institutional barriers that operate at the European as well as national levels. It is not possible to point to any great achievements so far.

The adoption of the Fifth Action Programme and its review process have coincided with the introduction of the doctrine of subsidiarity in the Maastricht Treaty (1993) and the measures taken to implement it at the Edinburgh European Council in December 1992. While some hoped and others feared that the effect might be to propose the rolling back of significant swathes of existing environmental legislation, such as the Bathing and Drinking Waters Directives (Hey and Brendle 1994), the Commission rapidly concluded that it should not venture into this territory, not least because of likely resistance from the European Parliament. What seemed to be the consequence, which was entirely welcome to the UK, was a noticeable reduction in the number of proposals coming forward to the Council and a more discriminating approach to their content, having regard to the principle of delegating decision making to the lowest appropriate level.

EUROPEAN PROFESSIONALISM INITIATIVE IN DoE

Under some recent Secretaries of State, the DoE has undoubtedly attached greater weight than it did in the 1980s to achieving positive results in substantive environmental negotiations in Europe. Such results, however, are not always easy to pinpoint and commentators still have their doubts about Britain's claim to be a leading environmental player in the EU (see Chapter 1). What cannot be contested, though, are the express steps taken by the Department since 1993 to improve its professional capacity for operating effectively in Brussels.

The text for this initiative is set out in a minute of August 1993 to the Prime Minister from John Gummer, then recently installed as Secretary of State for the Environment. The minute is reproduced in a departmental document which has been made available publicly. 'We have done much to change the culture of Whitehall in recent years', Mr Gummer wrote, adding:

> But there is one area where we need to do more. We need to encourage all Departments to be expert in their dealings with the rest of

Europe ... Every Department needs to keep in touch with the earliest stages of policy formulation affecting its business, long before the Commission puts forward proposals. We will be in a much better position to influence developments in Europe if we are present at the first stirrings of a new thought, rather than trying to retrieve the position later on by rational argument and diplomatic tactics ... Negotiating within the Community is a skill that can be taught. Officials dealing with Brussels should as a matter of routine get to know their opposite numbers around Europe before they become engaged in negotiations.

(DoE 1993b)

Following this minute a review of the DoE's handling of European business was conducted. The resulting report (*ibid*) became the basis for a programme of European professionalism across the Department, conducted under the auspices of the European co-ordinating Division in the Environment Protection command, which gave the Division an explicit role for the Department as a whole, not just those Divisions dealing with DGXI.

Although the burden of Mr Gummer's minute to the Prime Minister was that the environment side of the DoE was well versed in European ways, the rest of the Department needed to be brought up to speed. The professionalism initiative has helped to sharpen up the consciousness (and hopefully the performance) of environmental officials in European matters. The programme got under way early in 1995 with a high-level seminar on the European Parliament and its role following the Maastricht Treaty. John Gummer gave a keynote address, nailing his European colours firmly to the mast for the benefit of senior and middle-ranking staff from the full range of the Department's responsibilities. Apart from its symbolic value within the Department, this occasion paved the way for a more structured relationship between departmental officials and the deliberations of the European Parliament's Environment Committee. Hitherto the Department had tended to confine its contacts to standard briefings on proposals before the Committee and more tailored advice to individual MEPs who requested it. In view of the Parliament's increased powers following the Maastricht Treaty, this closer and more systematic involvement represents a prudent step.

The main outputs from the professionalism programme include the following.

First, a rigorous survey was conducted of all staff at Executive Officer level and above to establish their experience in European work, the training, if any, they had received, and their EU language competence and training.

Second, a policy note was prepared on the importance of establishing multi-skilled teams to see through all the processes of individual pieces of EU legislation from their conception within the Commission through to final adoption by the Council, with formal requirements to have Ministers approve a strategy at the outset and to involve all relevant Departments. The principle behind the note is that such legislation, in view of its precedence over domestic laws, should be taken at least as seriously as legislation going through the Westminster Parliament, to which interdepartmental rules and departmental practices accord the highest priority.

Third, a much greater emphasis has been placed on the legal nature of EU instruments, calling for negotiators to have regard to the feasibility and nature of UK implementing legislation from the outset.

Fourth, a more strategic and policy-driven approach has been adopted to the placing of departmental secondees within the Commission, as detached national experts, or in the institutions capable of yielding European experience. Hitherto a few policy directorates had established close enough links with Commission counterparts to be able to field UK personnel to fill gaps where new draft proposals of interest to us were on the agenda (as in the case of IPPC mentioned above), but generally secondees were pursuing personal career-broadening experience.

Fifth, from time to time seminars for departmental officials are held on non-legislative matters. Topics that have been covered include the European Parliament, as already mentioned, the operation of the European Regional Development Fund (ERDF) and European finance issues, such as the attribution to departmental budgets of EU expenditure decisions.

Sixth, a handbook for DoE officials on negotiating in the EU has been prepared, written by James Humphreys in 1995, recently returned from a period as Second Secretary in UKRep. Amusingly illustrated and in paperback format, it runs to 250 pages and covers everything from how to reply to an infraction letter from the Commission to advice on the best restaurants for diplomatic entertaining or an evening out on a civil servant's meagre allowance.

Finally, another recent feature which recognises the need to work closely with other member states, mentioned in Mr Gummer's minute, is the strengthening of links with the environmental attachés in the UK embassies in member states. The Department now arranges annual meetings in London with the attachés to explain in more detail UK policy developments and priorities and to discuss how to present them in host countries in a way that will support our negotiating stance in the Council. Or, to put it another way, how to bury once and for all the 'Dirty Man of Europe' image which can so easily be deployed against the UK when our good faith is in question.

CONCLUSIONS

There is plenty of evidence for the view that Britain's membership of the European Union has had a positive effect on its environmental policy and caused it to adopt tougher measures sooner than it would have done if left to itself. This view is also expressed in an article written by Derek Osborn, when he was Director-General, Environmental Protection at the DoE (1992). He analyses the contrast between the traditional UK environmental quality objectives approach and the Continental preference for the system of uniform emission limits. The positive effect of EU membership is particularly true of the area of water policy, where Britain had become complacent in imagining that its standards were among the most advanced internationally and that Britain's Continental neighbours, especially the southerners, lagged far behind. It is also true of nature conservation legislation where the momentum of some of Britain's earlier site protection measures had not been sustained. Where Britain has taken an environmental lead, for example on global warming, it has been helped by changes in energy policy made for non-environmental reasons.

Another important factor has been the nature of European Community legislation. Traditionally UK legislation has placed duties on UK subjects, often exempting the Government and its agencies, with scope for flexibility in the hands of Ministers in specific cases. EC legislation is quite different. It imposes obligations on the UK Government and enables the Commission to pursue the Government for alleged infractions all the way to the European Court of Justice. The main instrument of legislation, the Directive, has turned out to require much more precise implementation than appeared to be the case in 1972. Such legislation, while uncomfortable and sometimes uneven in its application, is generally positive for regulatory regimes and those who enforce them. In the environmental sphere the DoE can now argue, with the support of the Foreign and Commonwealth Office and the Whitehall legal network, that particular development projects will risk an adverse judgment in the European Court and should not therefore be pursued. Without the EC Treaties the environmental objection might more easily be brushed aside.

Arguably the UK is now having considerable success in encouraging a more strategic approach to environmental policy and legislation on the part of the Commission and the agenda for a sustainable development policy at the European level is beginning to be established. Under Euro-enthusiastic ministerial leadership DoE officials have adopted a more professional approach to negotiating and working with EU partners, thus learning by deliberate measures what others appear largely to take for granted. This is all the more commendable because, under the Conservatives at least, it had to be operated within the Government's cautious, if not negative, attitude to EU law making generally. The progress achieved, moreover, was against a background of successive staff reductions applied to the Department. Forging pre-negotiation contacts with other member states and the Commission and keeping other Departments, including the territorials, fully in the picture requires time and effort. The value of informal senior level meetings of Environmental Directors-General or Water Directors, which help to make Commission proposals and enforcement action realistic and balanced, could easily be lost if UK officials were too stretched to participate.

Where does this place the UK among its European partners as an environmental operator? In terms of its role in some of the global conventions such as those on climate change, the ozone layer and biodiversity, the UK is acting as a leading force for progress, firm targets and transparency. It has also set in train a substantial process within the UK to follow up the Earth Summit and work out its implications, putting it ahead of most other developed countries. However, real world environmental performance by the UK is like the curate's egg, as the OECD Performance Review of 1994 demonstrated. It would be very hard to show that, across the environmental sectors generally, the UK was leading countries such as Denmark, the Netherlands and Germany or the newest member states, Austria, Sweden and Finland. Rather, as Haigh and Lanigan (1995) have suggested, the UK is so far in a middle group of 'pragmatic' countries keen to ensure that environmental measures take full account of other interests and are not unreasonably expensive. This finding has been rather galling to some senior people in the DoE who feel fully committed to the environmental agenda themselves and have made a huge effort to bring along the rest of Whitehall. Nevertheless, the fact that the UK's most regular and substantial bilateral contacts on EU environmental issues are with France offers some confirmation of the Haigh thesis. This does not mean that blocs of countries with more or less advanced positions form a common feature of environmental negotiations. In fact linkage of separate issues is rather unusual.

The fact remains, however, that, in spite of strong environmental groups in the UK, at its heart the British establishment is not very green, as is reflected in DoE's pecking order among Whitehall Departments. There are many environmental assumptions, right or wrong, that have to be argued within the UK Government machine that simply appear to be self-evident in the German or Dutch context. When one discovers that the Danish Environment Minister is seen as the chief political rival to the Danish Prime Minister or that the Danish Environment Ministry was voted its own overseas aid programme by the Parliament because it is more trusted for environmental projects overseas than the Overseas Aid Ministry, some measure of the difference becomes clear.

One thesis of this chapter is that ministerial attitudes at the head of the Department of the Environment have played a significant part in determining the UK's stance in relation to developing EU environmental policy. While the underlying agenda may be influenced by many other factors, domestic and external, the Secretary of State for the Environment has the ability to apply the brake or the accelerator to the way the Department conducts its EU business and to the position taken on specific proposals. That this should be so is unremarkable and reflects the constitutional position in a democracy. It is also true that negative perceptions on the part of

Britain's EU partners take a long time to dispel, and for a change of line to be acknowledged a sustained effort at both ministerial and official level is required.

While the UK has made up a lot of ground as an environmental operator in the Union during the 1990s, its scope for becoming a standard setter is still limited. In this context it is likely to be most effective in pressing for a strategic approach, clear scientific underpinning and monitoring, enforcement carried out on a similar basis throughout the Union, cost-effective and integrated measures, and an agenda engaging the transport, agriculture, fisheries and energy sectors in sustainable policies. Of course some of these concerns could easily become a cloak for holding up progress and, if it is to be successful in its current proactive role, the UK will continue to find it necessary to demonstrate good faith by ensuring that, where valid comparisons, such as the OECD review process, show UK environmental performance to be wanting, the appropriate efforts are made to catch up.

NOTES

1 I was Director of Rural Affairs 1991–4 and Director, Global Environment 1994–5 at the DoE. The chapter was finalised in 1996.
2 LIFE is a financial instrument for the environment established by Regulation 1973/92 to assist the development and implementations of the EC's environmental policy. The acronym stands for L'Instrument Financier pour l'Environnement.
3 UKRep is the shorthand for the UK Permanent Representation to the European Communities, that is, the diplomatic mission to Brussels for the purposes of the UK's business with the organs of the EU. For officials working on EU issues it is a key channel since it provides the formal link with the Community institutions and is involved in all negotiations. It is an organisation led by the Foreign and Commonwealth Office and acts in principle as any other embassy. The DoE's relations with it are very highly developed and involve a daily to and fro of meetings, telegrams, faxes and telephone calls. The environment brief in UKRep is shared by a Grade 5 who has other responsibilities and a full-time Grade 7 and HEO(D), the latter both secondees from DoE at the time of writing.
4 The Structural Funds are intended to promote the economic and social development of disadvantaged regions, sectors and groups within the European Community. They support training, job creation and infrastructural improvements. During the period 1994–9 the resources available amount to 141 billion Ecu.

3

BRITAIN AND THE EUROPEAN POLICY PROCESS

Nicholas Hanley

The interface between the UK and Europe in the area of environmental policy is complex, problematic and dynamic. It is, however, possible to make some useful conclusions as to the impact of UK membership of the Community in this area and how the relationship continues to develop.

Although the UK, when joining the Community, had a fairly well developed tradition of national action on environmental protection and planning, and certainly more developed than the majority of other member states, there can be little question that, in many areas, the Community has forced the pace of action in the UK. In areas as diverse as bathing water standards and large combustion plant emissions, there can be little doubt that the UK would not have independently adopted the measures required by the Community (see Chapters 10 and 14).

Despite a generally good record of implementing Community environmental law, the UK has frequently, much to the annoyance of the Government, earned the label of 'Dirty Man of Europe' (Rose 1990). Compared to the performance of many other member states, this label is certainly not justified, but it has been to some extent self-inflicted by the handling of a number of cases where the UK Government has either opposed new initiatives, or sought to limit the application of particular Directives. A classic early case of this was the Bathing Water Directive, which imposed high standards on designated bathing waters. In an attempt to limit the potential cost of implementing the Directive, the Government designated only the twenty-seven waters which already met the standards laid down, implying in the process that the beaches of many seaside resorts were not bathing waters. The subsequent outcry by both environmental and tourist interests led to a change of policy and a significant investment programme. The attitudes which led to situations such as this were, ironically, perhaps generated by a sense of superiority induced by the very strength of the environmental management tradition in the UK (see Chapter 1). Some Directives in the early days were seen to be of little relevance for the UK but intended to

57

sort out the mess elsewhere in Europe. While this mentality no longer prevails, there is nevertheless a strong residual feeling that a level playing-field in environmental management is still a long way off, combined with a perception that the UK is unfairly treated by excessive attention to minor cases of poor implementation compared to other member states.

The attention given to cases such as these is, without question, due to the extensive nature and organisation of environment groups in the UK who have made ample use of the opportunity created by Britain's membership of the Community to make appeals to a 'higher authority' when not satisfied with the performance of the UK authorities. This leads us to the first important question, therefore, in analysing the relationship between Britain and Europe: who are the British?

BRITISH ACTORS AND EU ENVIRONMENTAL POLICY

A reflection on the above question leads to the identification of a wide range of British 'actors' on the scene of Community environmental policy making:

- *The voluntary organisations.* The UK has one of the strongest traditions of local and national environment movements in the Community. Their 'participation' in decision making in various ways is accepted at the national level and they have adapted well to the Community system, making full use of opportunities of complaint, petitioning, parliamentary lobbying, and so on. UK voluntary organisations have also played a major role in shaping the environmental lobby in Brussels, being active within the European Environment Bureau – the official umbrella organisation for the European environmental movement – but also instrumental in the creation of other more subject-specific groupings, such as BirdLife International, where the RSPB is a major player (see Chapter 6).
- *The local authorities.* The initial interest of local authorities in the European Community was lobbying for Structural Funds, but this has extended into wider policy areas, including the environment, where, in many cases, they will ultimately be responsible for implementation of EU Directives (see Chapter 8). Both individually, through the increasing number of representation offices, and collectively, via their national and international associations and representation on the Committee of the Regions, there is a growing local authority presence in Brussels. While still with limited influence in the legislative process, they are major players in taking forward policy initiatives such as the Green Paper on the Urban Environment (CEC 1990c), including notably the Sustainable

Cities Network, which brings together over 200 cities committed to implementing Local Agenda 21. This is, in fact, an interesting example of the divergence of UK views on particular initiatives. The UK Government has been generally reserved, if not indeed opposed, to Community initiatives on urban issues on the grounds of subsidiarity. Many UK local authorities have welcomed the positive lead taken and their representative on the subsequently created expert group, Colin Fudge, has chaired this group successfully for a number of years. The Local Government Management Board has also been very active in a number of the European networks and programmes which have been developed (Ward and Williams 1997).

- *UK companies and trade associations.*   There is a very large trade and business lobby in Brussels. British companies are prominently involved in representing their individual interests as well as being involved collectively in the many pan-European associations (see Chapter 9).

- *MEPs.*   The growing importance of the European Parliament, particularly since the co-decision provision of the Maastricht Treaty, gives MEPs a special role in the Community decision-making process. In the environmental field, British MEPs from both the Labour and Conservative Parties are prominent within the Parliament's Environment Committee. The Glasgow MEP, Ken Collins, current Chairman of the Committee, has, without question, developed a significant and influential role.

- *UK Commissioners and their Cabinets.*   The UK Government nominates two Commissioners (by convention, one from the Conservative Party and one from the Labour Party) and each Commissioner appoints a cabinet (private office). Each Commissioner has responsibility for a particular policy area, and between 1985 and 1989 one of the UK Commissioners, Stanley Clinton Davis, held the environment portfolio. However, the collegiate decision-making system within the Commission provides Commissioners, irrespective of their individual subject responsibilities, opportunity to influence any policy area. The preparation of decisions of the Commission via detailed examination of proposals in committees drawn from the various cabinets is a critical part of the process and frequently leads to modification or rejection of proposals.

- *British staff within the Environment Directorate (DGXI).*   Over the past few years, the UK Government has shown a growing interest in ensuring adequate recruitment of UK nationals into the Commission, particularly at the professional level (see Chapter 2). In addition to those in the recognised senior positions in the hierarchy (heads of unit, for example) there are a number of British staff at the level of file officer whose potential influence as the initial drafters of new proposals should not be ignored (Hull 1993). Reliance on detached national experts for technical expertise in many areas, which is particularly marked in DGXI, provides member states with a fairly direct way of ensuring that at least their way of thinking is understood within the service. In recent years, the UK Government has become far more positive about detachment of experts to DGXI, and such a secondment is now more clearly recognised as a valuable career stop, whereas in the past some returning staff had the impression that it was not regarded as of any particular value (Christophe 1993).

* *The UK Government.* Ultimately, via the Council of Ministers, national governments exercise a control over the development of Community policies, legislative proposals and budget decisions. Routinely, via the Council of Permanent Representatives (COREPER), Council working groups, management committees and expert groups, they influence, at official level, much of the day-to-day business of the Community.

This far from complete listing demonstrates the complexity of British influence on Community processes. On any particular issue, coalitions will exist between different actors, but rarely will all British interests be speaking with one voice. There may also be different perspectives on the desirable mode of Community action.

Traditionally, the main focus of environmental policy has been legislation and any retrospective analysis of influence would rightly focus on that process, but this pattern is changing. The Community's Fifth Action Programme (CEC 1992b) emphasises the development of new policy instruments and the integration of environmental considerations more fundamentally into other Community policies such as agriculture, transport, and industry.

This is an approach largely welcomed by a range of UK interests, albeit with different emphases. The environmental groups, for example, have frequently attacked the perceived misuse of Community regional funds for causing damage to the environment (Long 1995). Likewise, the UK Government has been in the forefront of calls for reform of the Common Agricultural Policy (CAP) and has pioneered the early adoption of such agri-environmental measures as environmentally sensitive areas (ESAs). However, the environmental groups also favour the development of a strong Community legislative framework for environmental protection (see Chapter 6), whereas the UK Government has stressed national responsibilities and the scope for flexible implementation by member states (see Chapters 2 and 10).

OPPORTUNITIES, ACTORS AND THE EU POLICY PROCESS

A useful framework for analysis of the ways in which policies are influenced is to consider the stages in the process and how different groups use the opportunities available at these different stages. Essentially, the process can be defined as involving: agenda setting; formulation of proposals; decision making; and implementation.

Setting the agenda

The process of setting the 'agenda' for Community environment action is, perhaps, the most complex to analyse and, undoubtedly, the most open (Peters 1994). Although the Commission, by the terms of the Treaty, has the sole right of initiative in respect of making formal proposals, the processes whereby subjects are recognised as meriting attention involve a combination of pressures from a variety of directions, including:

- *The example of action, or threat of action, by a member state or third country.* For example, the catalytic converter technology for emission control of vehicles had been in operation in the USA and other countries for many years before being incorporated into Community standards. Its adoption was brought about through pressure from environmental groups who were able to point to this third-country experience in persuading both the Parliament and the Commission of the technology's validity, against the arguments of the car industry (Arp 1993).
- *Council or Parliament resolutions.* Although not able formally to initiate proposals, both institutions can, and often do pass resolutions requesting the Commission to come forward with new proposals. These avenues have been used recently by, *inter alia*, UK interests in the animal welfare area pushing the Commission to respond to their concerns, and the UK Government keen to launch new initiatives on the enforcement and implementation of Community legislation.
- *Awareness raising through research, conferences, and the like.* These various tools can be very influential in stimulating debate on particular issues. Environmental groups are very active in this regard, but member states are also involved. The UK Government, for example, organised in 1996 an international conference on ozone pollution in cities, which called on the Community to advance proposals to tackle the sources of this particular problem.
- *Implementing international commitments.* Involvement in international activities has become of growing importance, with many Community proposals stemming from agreements made in various international fora (Haigh 1992). So the question of who represents the Community in such

fora has become more critical. Depending on the topic under discussion, there is a spectrum of situations, ranging from the exercise of Community powers (with the Commission as the key negotiator), to Community co-ordination (with the Council Presidency playing a key role), and to national competences (with each member state negotiating separately). However, even the most hesitant member states have increasingly recognised the value of Community co-ordination, providing as far as possible a united voice for the Community as a very influential and often decisive contribution to such debates. John Gummer, when UK Environment Minister, played a leading role, often in combination with other Community ministers, such as Sven Auken from Denmark and Klaus Topfer, the previous German Environment Minister, in achieving worthwhile international agreements.

- *Draft proposals.* A method sometimes used to 'help' the Commission in advancing new ideas is the submission of full draft texts for it to adopt. This has usually been done by environmental groups, but also by some member states and even occasionally by industry groupings. Contrary to general belief, industrial interests are not always against new legislation and have been known to request the Community to legislate in certain areas to prevent a proliferation of national measures likely to disrupt the single market in products.

The methods discussed above relate mainly to those for bringing forward new issues for action. Alongside this, there is a parallel activity of lobbying and persuasion by interested groups equally keen to keep issues off the agenda. This type of activity, as with the positive agenda setting, continues into the process of influencing the preparation of particular proposals.

Influencing proposals in the making

While the classic view is that individual officers drafting proposals can have a major influence on the shape of a finally agreed proposal (Hull 1993), in my view this is a situation that is rapidly changing. As the Commission moves towards a leaner legislative programme, individual proposals are increasingly subject to broader scrutiny, both within individual Directorates and in the interservice process prior to the submission of proposals to the full Commission. This formulative stage is, nevertheless, probably the most critical in the whole process, as is attested by the attention paid, particularly by industrial lobbies, to cultivating and maintaining contacts in the Directorates of interest to them.

The exclusive right of initiative of the Commission means that the best place to influence, kill or delay a proposal is within the Commission. The member states, including the UK, have become more astute in operating at this stage, in seeking to ensure that they are not confronted in Council with proposals they do not like. The tactics available to them include the formal opportunities of participation in expert groups assisting the Commission in preparing proposals, as well as informal lobbying of both the originating Directorates and cabinet and potentially opposing ones. The UK Government has, undoubtedly, played a significant role in delaying and modifying proposals for strategic environmental assessment in this way.

This stage of the process does, however, require detailed input of external expertise – Commission officials could often not proceed without it – so in no sense are all contacts at this stage to be characterised as offensive lobbying. The comparative lack of resources and of detailed technical knowledge of the environmental groups restricts their ability to operate effectively at this stage, even though they are now regularly invited to participate in expert group meetings (Mazey and Richardson 1992b).

In my experience, an example of the most sophisticated lobbying at this stage is that of the oil industry, which follows a strategy with two main components. CONCAWE, the Oil Companies' European Organisation for Environmental and Health Protection, is used to making detailed inputs aimed at demonstrating that: either there is no problem; the problem can be better solved by action in another area (for example, car emissions rather than fuel quality); there are more cost-effective ways of tackling the problem; or the proposed adaptation will require more time. Alongside this is the more direct political lobbying of EUROPIA (the European Petroleum Industries' Association) which, following on from CONCAWE, will aggressively lobby up to Commission level to kill or delay proposals that it deems still unacceptable. The individual companies in this industry co-ordinate their lobbying effectively, with each national company targeting its respective national Commissioners, and they rarely break ranks.

Decision making

Once a proposal has been made by the Commission, it passes to the Parliament and Council for final decision. Although there are examples of proposals which have failed to pass this stage, it is generally recognised that the exercise of influence at this stage is more likely to result in modification to proposals rather than defeating them outright.

The co-decision provisions of the Maastricht Treaty have given the Parliament significant, although ultimately largely negative, powers in respect to proposals under the Single Market articles of the Treaty (Wilkinson 1993). In the environmental field, where product standards are usually brought forward on this legal basis, these powers are therefore important. As a general rule, the Parliament, very much influenced by environmental groups, has pushed for more ambitious standards and deadlines for their introduction. Even here, we must be careful in generalising. In a recent case concerning motorbike noise limits, the Parliament was persuaded by a strong biker lobby (co-ordinated by British bike interests) to modify upwards a Commission proposed noise limit. It is certainly the case that, in the past few years, the lobbying of Parliament has become more intense, leading recently to the adoption by the Parliament of guidelines on this activity (McLaughlin and Greenwood 1995).

Within the Council, it is more difficult to identify any clear trends. Traditionally, the Germans, Danes and Dutch have pushed for higher standards and have frequently been isolated under majority voting rules but, in general, each issue tends to result in a specific set of coalitions and alliances. With the exception of certain highly charged topics, such as the CO_2 Tax where national fiscal independence was an issue, the position of the UK in recent years in environmental policy has not been characterised by the negative attitudes seen in policy areas such as social policy. Early examples where the UK delegation preferred isolation to compromise, usually at the insistence of Whitehall departments such as the Department of Trade and Industry (DTI), are now rarely repeated. UK officials spend a lot of time in Brussels negotiating in the working groups, preparing for Council meetings and, for the time being at least, this more positive approach guided by the Permanent Representation (UKRep) is seen to produce results.

Implementing

Although traditionally seen as the least interesting stage of the policy process at least politically, in recent years there has been growing attention to the deficiencies in the implementation of agreed Community legislation. Whereas the Commission services are able to follow closely the transposition of Community law into national provisions, it is less easy to follow the actual implementation of measures on the ground.

In principle, the procedures for reporting on implementation of Directives should be providing the Commission with an accurate picture of the state of play, but the inadequacy of the reporting provisions and lack of respect for those which existed was highlighted in a report to the Parliament by Caroline Jackson, the British Conservative MEP. An overhaul and standardisation of reporting has subsequently been introduced, but this will inevitably take some time to yield results. At present, much of the information coming to the Commission is by way of complaints by members of the public and environmental groups. The Commission is obliged to investigate such complaints and, where deficiencies in implementation are identified, to request action by member states and ultimately, in the absence of such action, to initiate court action.

In the UK the national groups and a myriad of local groups on specific issues generate a major case-load of complaints which is almost certainly disproportionate in terms of the UK's real performance compared to other member states. This has led the UK Government to join with the

Plate 3.1 Power lines in North Yorkshire. An environmental assessment procedure for a power station on Teesside which was approved by the Secretary of State for Energy in 1990 failed to take into account the effects of the associated power lines. The Council for the Protection of Rural England (CPRE) complained to the European Commission that this breached the Environmental Assessment Directive. The Commission agreed with the principle that such associated developments, if likely to have a significant impact on the environment, should be part of the overall assessment. However, it also accepted that in this case the required power lines would not have significant environmental impact.

Source. CPRE/Brabbs

environmental groups and the Parliament in calling for a less *ad hoc* system of enforcement. The idea of a Community Inspectorate has been raised, or at least an inspectorate of national inspectorates. The problem here is that not all member states have inspectorates. In this area, where the UK Government feels that its own performance is at a high level (with the example of the new Environment Agency being seen as a potential model), it is keen to push for higher Community standards, particularly if this can deflect criticism of poor implementation to those it is felt more rightly deserve it. Other member states are more reserved, singing the traditional British theme of national sovereignty to avoid interference of the Community in 'detailed national affairs'.

CONCLUSIONS

From the discussion above, some insight can be gained into the process of Community policy making and the different ways in which British interests have influenced it and continue to do so. Despite the common perception in the British media that Community affairs are dictated by faceless bureaucrats in Brussels, quite clearly the process is very open, with a balance of powers between the different institutions which rarely permits any particular group or interest to get things all their own way – the absence of a ruling party means that the outcome of individual issues is less predictable and genuinely open to wider interests to influence. The difficulty in this chapter of identifying areas where British interests have been more or less influential arises because almost all British actors on this scene have learnt the lesson that they will be more influential when part of a Community-wide grouping. This applies to the voluntary organisations, business interests and political parties alike and perhaps indicates that Britain is at last really joining the European Community.

4

REFLECTIONS ACROSS THE CHANNEL

Britain, France and the Europeanisation of national environmental policy[1]

Henry Buller

That Britain's environmental policy is being increasingly defined and transformed by European Union priorities, policies and implementation procedures is now little in doubt. A second-generation EU nation in an extended family that has now seen four periods of enlargement, Britain entered the Union at the time of the inception of EU environmental policy and its gradual integration into national law and practice. Given Britain's long-standing commitment to environmental concerns, the process of Europeanisation – involving not only the adaptation of national policy styles and agendas to European prerogatives but also their gradual convergence and, potentially, their eventual replacement by a common European form – was always going to have important consequences for the British way of going about things.

However, by definition, Europeanisation is an EU-wide process, affecting, to a greater or lesser extent, all the member states of the Union. To understand its impact in any one state such as the UK, we also need to examine its effects elsewhere. As Muller (1994: 63) points out, this leads to a paradox; on the one hand, Europeanisation reinforces the need for comparative analyses yet, on the other hand, ultimately renders them increasingly redundant. Clearly, Britain is not alone in feeling the growing presence of the EU within its body politic. It may give expression to that feeling more vociferously than others, for reasons we shall explore below, but Europeanisation is, surely, along with economic globalisation, the dominant force in the *fin de siècle* political evolution of all European states.

A comparative approach helps us to refine our understanding of domestic policy shifts by raising three central questions. First, how does the British experience compare with that of other EU member states? Second, to what extent can we throw new light on the British experience by approaching it from a different (in this case adopted Gallic) angle? Third, to what degree is the process of Europeanisation provoking similar institutional and/or policy changes, or raising similar issues, across national borders?

In seeking to provide some answers to these questions, this chapter compares the Europeanisation of domestic environmental policy in Britain with that in France and, in doing so, identifies a series of common Europeanisation issues. Despite their very different cultural and administrative contexts, certain parallel and indeed converging trends are discernible in the ways in which these two nations have 'responded' to Europeanisation in the environmental domain. The extent to which these responses are the result of intended policy or institutional harmonisation is, however, open to debate. Rather, they reflect a series of 'best-fit' responses which remain fundamentally conditioned by essentially domestic concerns and traditions. Thus, while at one level, the process of Europeanisation engenders, whether intentionally or otherwise, a certain rapprochement of institutional models and policy styles, national and local contexts remain, for the time being, the defining feature of environmental policy style.

TWO NATIONS, ONE EUROPE

Britain and France provide fertile ground for comparative analysis. Both were prominent players in post-war European construction, the European Community vision owing as much to Churchill's 'United States of Europe' concept as to de Gaulle's later 'Union of Nation States'.[2] Yet, from that early moment on, Britain has been the archetypal 'awkward partner' within Europe; being economically and ideologically closer to the new Anglo-Saxon heartland of the USA and culturally more at home with its own Commonwealth than with the Continent (see Chapter 1). Combined with a mistrust of the French that seemingly transcends class and political colour[3] and a widespread apprehension of German dominance in the EU, such independence, reinforced by concern for parliamentary sovereignty, has ensured that Britain remains a major bastion of resistance to the wholesale embrace of the European idea. By way of contrast, France presents a more paradoxical image. Being, to a certain extent, the classic nation state where republicanism and patriotism unite, France has been at the forefront of European construction, yet has been able, since the creation of the European Economic Community, to use European supra-nationalism chiefly to reinforce its own domestic strengths and individuality.

Significantly however, both the pace of European integration (including the enlargement of the Union) and the direction that integration is taking, are prompting broadly similar hesitations in both nations; though here the

French political elite is notably less divided than its British counterpart.[4] Three particular areas of popular anxiety might be identified. First, European integration is seen as being too technocratic, key decisions being made less by directly elected representatives than by a series of technical corps who, though answerable to the Council of Ministers and to a lesser extent to the European Parliament, none the less appear to critics to be acting independently of the stated wishes of individual member states. Second, Europeanisation is closely associated in the minds of many with a free market liberalism that not only implicitly challenges the individual interests of member states and their public service traditions but is also increasingly being seen as contrary to the original intentions of the former EEC. Finally, the broad 'method' of Europeanisation, one that places greater and greater reliance upon normative definitions and indicators, appears to be reducing culturally and historically defined national identities and practices to the level of variations in degrees of harmonisation to a hypothesised European standard.

Such reservations are gaining an increasingly high profile within the two nations under study here as the timetable for monetary union (EMU) approaches. In the UK, Euroscepticism, which has been a prominent component of both Left and Right debates since Britain's entry in 1973, has been fuelled by the prospect of EMU. However, in France too, commitment to the EU (and particularly to the EMU) is far from universal. The 1991 French referendum on the Maastricht Treaty was 'won' on the slenderest of margins (51 per cent for, 49 per cent against), while the French Left were returned to power in 1997 on a wave of popular reaction to the pursuit of the economic stringency in preparation for entry into EMU at the expense of national employment and social reform goals. In France, as in Britain, Europe currently has a 'bad press', and numerous newspaper headlines have drawn attention to the average citizen's general wariness over extending further the influence of the EU into French daily life.

Environmental protection has arguably been one of the policy domains in which the EU has come furthest in dominating national agendas and styles. Most national environmental policy derives from legislation emanating from Brussels. Yet, while this is also undoubtedly true for some other sectors, particularly agriculture and trade, what distinguishes the environmental policy domain is, first, the fact that for certain member states the environment as a distinct public policy domain has been to a large degree created by European integration requirements and, second, the degree to which the European legislative regime has brought about or contributed to fundamental shifts in the way nations address environmental protection and management within their frontiers.

HENRY BULLER

DIFFERENT ENVIRONMENTAL POLICY
TRADITIONS

France and Britain display considerable variations in their environmental policy traditions. In part these derive from their very dissimilar institutional structures (see Table 4.1), the one founded upon a pragmatic approach to achieving effective solutions to individual issues, the other concerned with establishing clear legal frameworks and divisions of responsibility enabling the Republican state to meet all eventualities.

Yet these variations also result from the very different environmental policy contexts (see Table 4.2). France is one of the largest European states, whose territorial extent, on the one hand, and physical and ecological diversity, on the other hand, have historically been seen as an infinite richness to use and exploit rather than as a vanishing heritage to cherish and protect. This has given a distinctiveness to the development of French environmental concern which has been focused upon either designated sensitive zones or the largely technical issues associated with rather late urbanisation. Britain, on the other hand, is characterised by early urbanisation and industrialisation, necessitating pioneering forays into environmental health management, a threatened countryside, and a tradition of scientific, aesthetic and philanthropic concern for nature and access to it. Inevitably, therefore, environmental problems have been differently conceived. In France, for example, raw sewage has traditionally been allowed to disperse through septic tanks on land; whereas in Britain raw sewage is traditionally pumped, often via relatively short and fast-flowing rivers, out to sea. Likewise, in France, agriculture has long been regarded as essential both in maintaining the productivity of land and in keeping 'savage' nature at bay; whereas in Britain, an overriding concern for the aesthetics of landscape pervades the national territory, and farmers, once seen as guardians

Table 4.1 Some key differences between French and British institutional structures

	Britain	*France*
Some defining 'isms'	Pragmatism, Realism, Empiricism	Idealism, Humanism, Republicanism
Dominant juridical form	Primary legislation and jurisprudence	Constitutional law and codified legislation
Policy style	Good practice and negotiation, case by case	Legal and administrative procedures overseen by state bodies
Locus of action	Traditional importance given to local government	Primacy of the central state, dominant political elite

70

Table 4.2 Some key differences between the French and British environmental policy-making traditions

Britain	France
Long history of environmental concern (though restricted to certain areas, notably landscape and wildlife protection); internationally recognised environmental management expertise (largely in the above fields)	Short history of specific environmental concern (other than that associated with the management of natural resource use); long considered to be 'behind' other European states in the maintenance of environmental quality
Department of the Environment, one of the major Ministries of State, encompassing a wide range of competencies including local government and local government financing, planning, environmental protection, infrastructure and so on	Ministry of the Environment, a small ministerial body with a highly specific brief and relatively few independent policy domains; unlike other ministries, has no local level services of its own
Local government traditionally a central actor in environmental policy making and implementation with a major role in land-use planning and environmental health protection and, until recently, waste management and water planning	Central state traditionally the key actor in environmental policy making and implementation, political decentralisation being relatively recent and subject to central state mediation
Strong, well organised and engaged environmental and consumer lobby, ready to identify policy lacunae or blow the whistle on poor policy implementation	Emerging but as yet relatively weak environmental and consumer lobby notably less ready to take on state actors; strong and well established tradition of political ecology (Green parties) dating from the early 1970s and enjoying relatively high profile and occasional electoral successes

of nature, are now more often stigmatised as its despoilers (Lowe and Bodiguel 1990).

How then have these two nations, and their respective environmental policy traditions, differed in their responses to the growth and development of EU environmental policy? There seems little doubt that we have come a long way since Wallace (1972: 286) was able to claim that the EEC (which, at the time included only the original six nations) had not 'penetrated dramatically into the national political scene' of the member states. The bulk of domestic environmental policy making in both France and Britain today is essentially based upon EU legislation. If one takes the example of water quality protection, EU Directives have had the effect not only of creating a common environmental agenda but also, by establishing specific standards for certain water uses and imposing mandatory sampling and reporting procedures, of creating a context for a growing degree of similitude between national management procedures (Ward *et al*. 1995; Buller 1996a; Chapter 14).

However, rather than simply enumerate the multitudinous 'responses' to the Europeanisation of environmental policy visible in France and Britain, which would include institutional changes, shifts in policy style and modes of representation, changing environmental agendas and the introduction of new modes of environmental quality assessment and control (see for example, Buller *et al*. 1993; Ward *et al*. 1995; Buller 1996b; Jordan *et al*. 1996), the remainder of this chapter will focus on a set of common debates and issues, with the intention of highlighting both divergent and convergent trajectories in the French and British experiences of Europeanisation.

ENVIRONMENTAL SERVICES AND PRIVATISATION

Over the past fifteen years, we have witnessed a gradual erosion of the European tradition of public services operated by state or municipal employees for the purpose, not of generating wealth, but of providing for the public good. That erosion has occurred either through the privatisation of formerly public services or through the contracting out of certain functions to private companies. In both cases, private concerns, no longer fettered by the overriding principle of public service, seek to make a profit from the provision of necessary public goods. Environmental services in Britain and France, notably water supply, provision and treatment, have been particularly affected by the drive towards privatisation. That this has, in general, coincided with growing EU environmental legislation suggests that the two processes may be linked, either in a direct causal way or in the sense of both being responses to a broader dynamic. Has the Europeanisation of domestic policy thereby encouraged, either directly or indirectly, the privatisation of environmental services?

At first sight, Britain and France present two very distinct models. The total privatisation of regional water services in England and Wales in one fell swoop in 1987 stands in stark contrast to the delegated privatisation of municipal water services that has, in some French cities, been in place since the beginning of the century (Lorrain 1991). Nevertheless, the 1980s saw a significant rise in French water service privatisation. By 1993, around 80 per cent of the French population received water from private distributors against around 60 per cent in 1983 (Buller 1996b). Neither the French nor the British models can be said to have been the direct consequence of European developments. Rather they have emerged from essentially domestic trends, born either out of ideological commitment, as with the sell-off of public utilities under the Thatcher Government of the 1980s, or out of economic necessity, as individual French local authorities responded pragmatically to the increasingly high investment costs of drinking water distribution.

However, privatisation cannot be dissociated from Europeanisation. The latter process has reinforced the arguments for the former. The free market ideology of the Single European Market encourages the competitive entry of private capital into areas formerly reserved for state monopolies. At the same time, the imposition of mandatory water quality norms and standardised monitoring, control and reporting procedures under EU legislation has both provided the basis for the entry of private capital and unquestionably led to increases in the costs of clean water provision (as well as the technical expertise required of operators). Privatisation has thus emerged both as a means for government and local authorities to offset and redistribute the costs of implementation and conformity (and to overcome their own technical limitations) and as a means for operators to free themselves from limited state budgets. Indeed, it was the prospective investment requirements due to EU Directives that were the most important factor in the UK Government's decision to privatise the water industry (Maloney and Richardson 1995), as they were for many French local authorities. Yet neither in Britain nor in France have privatised water companies had to bear the full brunt of implementation costs. In Britain, the water services companies benefited from the so-called 'green dowry' offered to them by the government in 1989 whereas in France infrastructural grants are regularly made by state and local government bodies to individual local authorities.[5] Furthermore, charges to the consumer have risen sharply in both countries, a trend water companies on both sides of the Channel attribute almost exclusively to the necessary costs of meeting EU standards (Buller 1996b). As a result, private water companies have been able to record considerable profits and offer their senior management large wages; this, however, at the expense of public confidence and increasing dissatisfaction notably over failures in supply management and inexorably rising charges.

Arguably, too, the introduction of an EU legislative and normative regime guiding water sector investment and policy has facilitated privatisation by creating an alternative (and international) regulatory framework to that traditionally subsumed, at the individual nation level, by state-guaranteed notions of public interest and societal goods. It is notorious not only that the pre-privatisation Regional Water Authorities sometimes failed to provide water of a potable quality but that they were their own regulatory authority. Here there is a clear paradox, one that derives from a shifting interpretation of the public interest. It is no longer enough that public services be in public ownership, though this belief seemingly retains widespread support. They need to be capable of delivering sustainable environmental goods of a required quality at a reasonable price and be transparent in both their achievements and their failures. The EU regulatory regime is increasingly providing a framework for these latter requirements, thereby rendering public ownership *per se* unnecessary. The critical difference between nation states thereby lies in the form that privatisation might take and the degree to which it remains under ultimate state control (Buller 1996b; Bodiguel and Buller 1996).

CENTRALISATION, DECENTRALISATION, REGIONALISATION: TOWARDS A NEW TERRITORIALITY

It has been argued that Europeanisation has had the effect of centralising environmental policy within Britain (Haigh 1986; Haigh and Lanigan 1995). Central government is the principal contact point with the European Commission and Council of Ministers, is required to transcribe EU rules into national legislation, and is thereby ultimately responsible for compliance failures. Local authorities, meanwhile, have gradually seen their own environmental powers reduced as regulatory functions have been regrouped within national and regional environmental agencies. In France, in contrast, there has been an emergence of local initiatives and powers in environmental management and policy making and this too has been associated with Europeanisation. How can we reconcile these two apparently contradictory trends and to what extent is Europeanisation actually the cause of either?

In seeking to respond to this question, we are immediately confronted with the difficulty of dissociating the impact of the EU from that of internal political shifts. Under successive Thatcher governments, political power in Britain was increasingly centralised while local government, the traditional home of British environmental planning and management, was consistently weakened. As was the case for privatisation, the driving force for change was not necessarily Europe-led but rather the result of a domestic agenda. In France, the critical political shift of the early 1980s was decentralisation and the transfer of former state competencies to local government. Any rise in local environmental initiatives needs therefore to be seen in this context. Local commune and *département* authorities today possess statutory roles in environmental policy making and implementation far in advance of the very limited capacities they had prior to 1982.

EU environmental policy has none the less fed into these internal trajectories, providing French local government with a new environmental agenda and reinforcing the powers of the newly centralised regulatory authorities in Britain. Here, France and Britain maintain their divergent paths. Yet, perhaps, the real significance of Europeanisation lies less in its impact upon the local and central levels than in its empowerment of a relatively new level of policy making, at least for Britain and France, that of the region. In both nations, regionally organised bodies, be they policy-making institutions, government offices, agencies or representative structures, have risen in response to European prerogatives.

In France, it has been the regional assemblies and state agencies that have proved among the most adaptable to European-led environmental concerns (Bodiguel and Buller 1994). In 1991 the regional bureaux of the French Environment Ministry, the DIREN, were accorded substantial new roles in the domains of water management and wildlife and habitat protection. Their equivalents in the agricultural policy sector, the DRAF, are emerging as the key promulgators of European-led agri-environmental policy in regions as diverse as Nord-Pas-de-Calais and Languedoc-Roussillon. Perhaps more significant, however, has been the increasing emphasis in France placed upon contractual planning and management. Here, the regions have become critical intermediaries between the central state and the European Union, in the one side, and local authorities on the other. Region–state contracts, essentially an initiative of the post-decentralisation era, have also become the basis for guiding environmental investment (that is, public investment in new environmental infrastructure, pollution control facilities, land rehabilitation, and so on): hopefully, they will foster a territorial coherence that has, so far, been the key failing of political decentralisation. Indeed, regional planning and the regional co-ordination of natural resource use and protection are emerging as critical themes for future environmental policy making across Europe.

In the broad European context, Britain remains disadvantaged by its 'administrative' rather than 'political' regionalisation. Nevertheless, the requirements of administering EU regional aid have helped catalyse a regionalisation of government structures. Regional offices of government in England, which bring together several national ministries, are now responsible, alongside the Scottish, Welsh and Northern Ireland Offices, for the co-ordinated delivery of a range of government programmes. One of the broad objectives of the regional offices is to promote the sustainable development of their regions, and it is at the regional scale that environmental actions are increasingly being promulgated and co-ordinated. The government regional offices have the responsibility for ensuring the integration of environmental considerations into economic development programmes. They have also been given a central role in promoting government policies for the countryside. Since the early 1990s, regional planning guidance has been given a strengthened role in the co-ordination of land-use planning, including the promotion of policies to curb transport growth and protect rural land. The new Environment Agency has been set up with a regionally decentralised structure (and with a separate agency for Scotland). At the same time, EU agri-environment and rural development policies have obliged the Ministry of

Agriculture (which remains detached from the regional offices) to relate its promotion of the farming sector more closely to different local and regional circumstances. The countryside conservation agencies have responded by emphasising the regionally differentiated nature of the rural environment. A similar trend is observable in France, where EU agri-environment policy has been interpreted as a major element in the move away from a central mono-lithic and state-led agricultural production policy towards a greater hetero-geneity of local agricultural-environmental management systems (Boisson and Buller 1996).

At the risk of seeming to pull a rabbit out of a hat, some of the apparent contradictions outlined above can now be resolved. To a nation like Britain, with a historically strong local environmental policy tradition, the growing role of public and private regionally organised bodies might be interpreted as a centralising tendency. To a nation such as France, where traditionally the central state has been the key player in environmental policy making, European-led regionalisation appears as a decentralising move.

If Europeanisation has led to movements in the vertical distribution of responsibilities within national power structures, a potentially more signifi-cant shift might ultimately be the development of more horizontal approaches to environmental management, ones based less upon adminis-trative and political divisions of national territory and more upon natural, ecological and hydrographical boundaries, which might, on occasion, tran-scend national frontiers. To a certain degree, elements of the regionalisation trend identified above have been in part predicated upon such principles. The regional organisation of the UK Environment Agency was inherited from the National Rivers Authority, whose regional structure, like that of the French *Agences de l'eau*, was based upon the supply rationale of large river basins. Likewise, agri-environmental measures under Regulation 2078/92 have targeted natural or semi-natural systems such as wetlands or bocage, and have been put together by member states in zonal programmes. Increas-ingly, the trend is towards a new territoriality in environmental policy making founded upon sustainability criteria, of which the European Union is a central advocate. Nation states, however, and in particular Britain and France, are likely to resist such developments. Not only do they go against established practices but they challenge existing sectoral and hierarchical divisions of competence.

NEW NATIONAL INSTITUTIONS

The establishment of the environment as an important field of domestic policy has coincided, in both countries, with the Europeanisation of that field. For example, the UK Department of the Environment was set up in 1970 by the recently elected Heath Government, through a merger of the existing Ministries of Housing and Local Government, Public Building and Works and Transport (subsequently, the Ministry of Transport was separated off). The French Ministry of the Environment was set up one year later, in 1971, by President Pompidou. Initially known as the Ministry for Nature Protection and the Environment, it was a new structure born largely out of the twin concerns of centralised regional planning and natural resource protection, previously the domains of the national planning agency, DATAR, and the Ministry of Agriculture (significantly, water quality protection, currently a central pillar of the Ministry's responsibilities, was only added to its list of competencies much later). The initial movement of the European Community into the environmental field can be seen as a response to such domestic initiatives. Since then, the French and British Governments have been confronted by essentially the same problem of policy and political management, namely how to establish on a firm footing this increasingly important policy field whose responsibilities often cut across those of more powerful policy sectors and where the locus of policy making was steadily shifting to Brussels.

Superficially at least, the response of the two governments has been quite different. The Department of the Environment is a large 'super-Ministry' with a broad range of responsibilities, most of which are non-environmental. It comes fairly high, though not in the very top rank of government departments. The French Ministry of the Environment, in contrast, is a very focused structure with relatively few independent policy-making powers. It acts largely in concert with other more powerful ministries. The two cases represent different resolutions of the dilemma of how to give both weight and focus to an emergent policy field within established government structures.

In the British case environmental responsibilities were attached to a big Department with some clout in Whitehall and the Cabinet. However, these responsibilities have always been a minor component of that Department's portfolio, which has also included responsibility for local government (the domain, in France, of the Interior Ministry) as well as for a number of functions actually antagonistic to the environment, including being the lead department for the construction and mineral industries.

The French Ministry has suffered no such lack of clarity of purpose and has always been an advocate of environmental protection. It often appears more as an internal government pressure group than the central focus of a major sectoral policy domain. In its early years there was a constant risk of it being marginalised. Indeed, between 1974 and 1984 it suffered no fewer than five shifts in its location within government, as it was successively tagged on to different portfolios ranging from 'Youth and Sport' to 'Culture' and, in 1984, 'Planning'. Only really since the late 1980s has the Ministry been able to consolidate its position within government. From the outset, its influence within the government machine depended largely upon the presidential patronage it enjoyed (favoured during the Pompidou and later Mitterrand years, neglected and, some would argue, actively weakened under Giscard d'Estaing).

In many respects, the equivalents within the British structure of the French Ministry of the Environment are the various environmental and conservation agencies, which likewise are focused, expert bodies that combine various technical and policy advisory functions. Though equally dependent on ministerial patronage, unlike the French Ministry the UK agencies operate at arm's length from central government and have responsibilities that are fragmented in terms of the kind of activity (responsible for landscape protection, nature conservation, pollution) as well as geographically (for England, Wales, Scotland and Northern Ireland).

How have these different structures then responded to the Europeanisation of the policy field? The growing significance of the environment on the European agenda has also raised its profile within national politics. This has contributed to an increase in the authority of the French Ministry while it has given greater prominence to environmental protection concerns within the Department of the Environment. The importance of the Environment Council within the affairs of the Community has also helped give the French Ministry some of the political clout it previously lacked while the growing input of European environmental legislation has furnished the Ministry with an increasingly important regulatory role which, combined with its widely regarded technical expertise, has been a significant factor in its gradual empowerment. Its leanness, focus and expertise have also enabled it to respond adroitly to the opportunities and requirements for lobbying and technical input thrown up by the Commission's policy making processes.

The growing salience and Europeanisation of the environment, however, have placed some evident strains on the British system of environmental policy making. In 1990, for example, the Secretary of State for the Environment, Chris Patten, revealed his frustration at being unable to deal with pressing environmental concerns (most of them with a strong European dimension), due to the ongoing political controversy surrounding the Poll Tax which also fell within his remit (Cole 1995). Nevertheless, there has been a steady build-up of the Department's policy making capacity on the environment (see Chapter 2). This has tended to eclipse the policy advisory role of the agencies. The growing volume of European Regulations has given a renewed but rather different emphasis to agency solutions, as the Government has looked for efficient means of policy delivery. This has resulted in the reorganisation and consolidation of the agencies (leading to the establishment, since the late 1980s of, *inter alia*, the Environment Agency, the Scottish Environmental Protection Agency, the Countryside Council for Wales, and Scottish Natural Heritage), but also the exercise of a tighter political leash over their activities. One consequence has been, at times, a somewhat uneasy tension between the Department and the agencies over the political and technical inputs into EU policy making. In such as English Nature, the Countryside Commission and the Environment Agency, Britain undoubtedly possesses some of the greatest concentrations of expertise in their respective fields of any country, but it is not clear that this is being deployed to the greatest effect within Europe (see Chapters 5 and 12).

Both the French and British Ministries have indubitably come of age and consolidated their environmental mandate with the growing importance of Europe as a source not only of environmental legislation but also of environmental policy innovation. Yet the current thrust of that policy, away from the sectoral concerns of the 1970s and 1980s towards a more holistic agenda, centred on integrated pollution control on the one hand and sustainable development on the other, has exposed the weaknesses as well as the strengths of the two models. Both remain limited by the entrenched sectoral constraints that are, to a large degree, inherent in governmental administration. At the time of its inception, the French Ministry was seen as a radical departure from the classic government Ministry representing a distinct set of interests and possessing its own corps of administrators and technicians. Whereas it has undoubtedly become adept at negotiation, consultation and co-ordination within the ministerial landscape of France, it nevertheless remains a relatively minor force, albeit one with a very clear sense of purpose. The British Department of the Environment, though it possesses the considerable weight of a major department of state, has not always been able to overcome similarly entrenched sectoral interests: certainly as evidenced by the government's environmental White Paper and Sustainability Strategy (UK Government 1990; 1994b). Both institutions are the product of their respective national politico-administrative and indeed environmental traditions (Buller 1996c). While the recent bouts of agency reform on both sides of the Channel have helped to create a greater focus for environmental concern and an internally more coherent environmental policy, it remains to be seen whether the consequences will be a stronger environmental voice within government or the marginalisation of environmental issues.

COPING WITH SOVEREIGNTY LOSS

There is little doubt that Europeanisation implies a reduction in the day-to-day policy independence of member states. Nations are no longer entirely free to set their own environmental standards or to pick and choose their own environmental priorities. How, then, have states dealt with this apparent loss of autonomy and is the British response in any way particular? The examination of European environmental policy integration reveals that apparent sovereignty loss can be and is compensated for in three ways; first, through the adoption of a national stance with respect to European legislation in general and environmental legislation in particular; second, through European diplomacy and representation; and third, through distinct implementation strategies that can have the effect of limiting or containing the domestic impact of EU legislation.

In Britain, one is frequently reminded of the impact of EU environmental legislation in the press and ministerial statements. Articles and comments decrying Brussels interference in traditional British practices are legion, as are examples of media outcry when Britain receives an unfavourable ruling from the European Court over failure to conform .

The British strategy has typically been one of 'externalisation', drawing attention both to the differences between EU rules and established domestic practices (and implicitly if not explicitly, the superiority of the latter) and to the negative financial and cultural impacts of EU legislation. The stance is largely critical in that it presents EU legislation as either unnecessary, too costly or founded on at best flimsy scientific evidence (see Chapter 1). The speed with which the British Government announced its success (which later turned out to be greatly exaggerated) in negotiating a relaxation of standards relating to drinking and bathing water quality in 1993 illustrates the extent to which such standards are portrayed as being wholly external to British interests. Ironically, however, this externalisation – not least because of its dishonesty in concealing the complicity of the British Government in EU decisions – has ultimately strengthened the hand of those pressure groups ready to expose British compliance failure.

In France, as indeed in certain other member states, the EU origins of environmental standards and rules are less immediately discernible. In contrast to Britain, France has generally adopted a policy we might label as 'internalisation', whereby EU legislation is transcribed into national law (as indeed it is in Britain) yet only becomes publicly visible once its French legal credentials are confirmed. State agencies rarely deal with Commission texts directly but await their incorporation into national law while individual lower tier local authorities are often unaware of the European origins of environmental rules. As a result, environmental interests frequently act in consort with state bodies to improve EU rules rather than seek to expose differences between national and EU positions.

There exists perhaps one glaring exception to this otherwise fairly universal picture, an exception that proves the rule. Of all the environmental policy domains in France, that relating to the agricultural environment has been arguably the most affected by the Europeanisation process, or so it would seem. While France, from the late 1960s onwards, repeatedly presented farm withdrawal and land abandonment as the principal 'environment' threats to the countryside, environmental groups and, to a certain degree, government bodies in Britain and certain other countries, such as the Netherlands and Denmark, stressed the impact of modern agricultural practices on wildlife, ecosystems and the rural landscape. The adoption of Regulation 797 in 1985 permitting the establishment and financing of Environmentally Sensitive Areas (ESAs) within member states reflected particular British preoccupations (Baldock and Lowe 1996). In France, ESAs were not only very late in emerging, but they were tellingly referred to as Article 19 zones (after the relevant passage in Regulation 797/85), thus emphasising their European origin (in a similar vein to British references to Euro-beaches and the like). In general, the gradual expansion of EU legislation targeted at the farming environment has taken the French agricultural policy community by surprise, leading not only to policy and institutional shifts within existing structures but also to struggles for legitimacy between the Ministries of Agriculture and the Environment (as indeed has happened in Britain). Undoubtedly, it is the way in which the issue is thus disturbing established hierarchies that has led some to emphasise the 'unFrenchness' of it all.

Both France and Britain are recognised as hard negotiators within the European Union's policy-making structure, though here again the relative positions of the two states within the European concert are apparent. For the most part, France has always appeared to enjoy an insider's position within Europe while Britain has persistently played the role of the outsider. Part of the explanation for this difference is undoubtedly historical, but it also reflects differences in the European agenda of the two states. While France's permanent representation at Brussels has traditionally been dominated by the Ministry of Foreign Affairs, the UK has traditionally adopted a far more sectoral approach to Brussels diplomacy (Lequesne 1993; Chapter 2), seeking specific policies and legislative changes rather than contributing to a broader European vision. However, one might argue that today such differences are being attenuated. Indeed, in the post-1991/92 climate, it is perhaps France, more than Britain, that is having to shift its European strategy (Prate 1995). In the environmental policy field, this has been reflected recently in a number of common challenges to proposed or existing standards (for example the lead standard in the proposed revisions of the Drinking Water Directive) while the growing importance of private environmental service providers is also a factor in the presentation of an increasingly shared position (Rousseau 1993). While the French retain their traditional reluctance to embark upon widescale deregulation particularly in the light of recent events in the public health domain, from the contaminated blood scandal of the late 1980s to the current BSE affair, the increasing parallelism of both the EU environmental agenda and the regulatory and institutional framework of implementation is, we might suggest, contributing to a convergence between British and French stances towards the Union.

Differences between the French and the British experiences are also revealed in implementation strategies, which are an important means of coping with sovereignty loss and of imposing national agendas on mandatory EU rules. Three relevant strategies might be identified; designation, measurement, and reporting. In their choice of designated sites for the application of European rules, states can seek to minimise the impact of EU legislation. Similarly, in the measurement of environmental quality parameters or in the reporting of results, states are able to influence the picture of conformity that is being given both to citizens and the European Commission. Thus Britain has been an assiduous user of designation strategies in, for example, the classification of bathing and shellfish waters (see Chapters 10 and 14). France, given the legal definition of both these waters, has found it more difficult to de-select controversial sites and has therefore relied more upon variations in measuring and reporting procedures to overcome local difficulties in conformity (Jordan et al. 1996).

Ultimately, the adoption of these various strategies reveals the current limitation of the Europeanisation process. National responses to Europeanisation have depended upon commitment not only to the European idea but also to the principles of sustainable environmental protection. To a certain degree, the two remain at odds despite the growing role of the European Union as the critical driving force in pushing the sustainable agenda. For member states, conformity to EU environmental legislation does not necessarily ensure the adoption of sustainable policies. At the onset of EU environmental policy, conformity was judged in terms of legal transposition (which is still the case for the more recent members), evaluations of the actual implementation of policy being set aside for a later date. As broad legal transposition has been achieved, conformity appraisal has shifted towards assessment of compliance to mandatory standards. But, in a number of policy fields, formal compliance is revealing itself as being ultimately environmentally damaging. For example, the designation of certain coastal bathing waters allows the continued degradation of undesignated (including inland) sites. Alternatively, by focusing on ground water sources for drinking water (and thereby ensuring a higher rate of compliance), states are able to ignore the deteriorating quality of surface waters. What is needed now is a third level of assessment, one that takes into consideration the real and sustainable impact of policy on environmental quality. Here Britain and France occupy a similar position. Conscientious implementors of EU policy, both remain pragmatic in their prioritisation of environmental imperatives. In the environmental sphere at least, neither has anything to fear from the other, though, following the recent adhesion of Austria, Finland and Sweden, three nations with rigorous environmental legislation, both might find this position ultimately challenged by the European Union.

NOTES

1 Research on which this article is based was financed by the Global Environmental Change Programme of the Economic and Social Research Council, award number L320273062, when the author held the post of Research Fellow at the Department of Geography, King's College London.

2 There is here a curious inversion. Churchill's apparent appeal to Euro-federalism and de Gaulle's defence of national sovereignty today seem almost the reverse of currently promoted Euro-stances of their respective nations.

3 See, for example, the diaries of the former Trade Minister Alan Clark (Clark 1993) or remarks made by the former Prime Minister James Callaghan to John Cole (Cole 1995).

4 The French political elite is remarkably homogenous when compared to its British counterpart. A large proportion of elected members (*deputés*) of the French Parliament on both the Left and the Right have been trained at one or other of the elite *grandes écoles* and there is no trade union sponsorship of left-wing members.

5 Under the French system, the infrastructure of water services remains in public ownership, private companies being effectively awarded licences to operate them and make profits from their use.

NOTES

Part III

INSTITUTIONAL
DYNAMICS: CHALLENGES
AND OPPORTUNITIES IN
THE EUROPEAN GAME

Whereas Part II took a primarily macro overview of Britain's environmental relations in Europe, Part III examines the impact of integration on specific institutions, namely: environmental pressure groups, local authorities, environmental agencies, and business and industrial lobbies. For each type of policy actor, the respective chapter covers the substantive impact of EU legislation on their roles; adaptive changes they have made in their procedures and practices; their evolving orientation towards, and links with, EU institutions; and the development of transnational approaches, perspectives and structures. This part of the book has three major aims. First, it seeks to identify which groups and institutions have benefited or been disadvantaged by European integration. Second, it uncovers the structural opportunities and hurdles to participation in the European game which help explain the differential learning curves for the various actors. Finally, the experiences of the organisations detailed here can be compared with those of central government, and should therefore act as a reminder not to characterise the impact of integration solely on the basis of government's relations with Europe.

Philip Lowe and Stephen Ward's chapter presents an overview of the response of three important types of policy actors – environmental groups, local authorities and environmental agencies – and analyses the overall restructuring of domestic roles and relations resulting from Europeanisation (Chapter 5).

Chapter 6 by Tony Long, the head of WWF's Brussels office, provides a more detailed account of the evolution of the Brussels environmental lobby and the involvement of UK environmental groups. Although Long agrees with the majority of commentators that UK environmental groups have generally adapted well, he emphasises that the environmental lobby remains underdeveloped and 'punches below its weight' at the European level compared with the national level.

In Chapter 7 Brian Wynne and Claire Waterton take a somewhat different approach by examining the establishment of an EU institution, the European Environment Agency, and its implications for member states and sub-national policy actors. The Agency is charged with assembling environmental information to support policy development and monitoring at the European level. The decisions it is taking on what counts as significant information and who should have access to it have profound implications for how environmental problems are publicly understood and how debate about them is structured.

The remaining chapters of Part III look at two increasingly important sets of actors in the European environmental arena with Janice Morphet, a chief executive of a local authority, writing on local government (Chapter 8) and Edwin Thairs, head of environment for the Water Services Association, writing on business and industrial lobbying (Chapter 9). Both sectors have had a long involvement in European lobbying but are relative newcomers to the European environmental arena. They illustrate how that arena is becoming ever more crowded and how the struggles between key domestic actors are being reconstituted at the European level.

5

DOMESTIC WINNERS AND LOSERS

Philip Lowe and Stephen Ward

European integration presents both opportunities and difficulties to nationally based organisations. Cumulatively, their differential responses and reactions effect changes in domestic political structures. Here we draw together the results of surveys we conducted in 1994–5 of UK environmental groups,[1] local authorities[2] and environmental agencies[3] concerning their orientation towards European policies and political processes on the environment. Our analysis seeks to identify the key opportunities and difficulties facing these organisations and the consequences of their differential responses and reactions for structural change in the environmental policy field.

European integration manifests itself at the domestic level as yet another factor that competing organisations seek to deploy to their advantage in their continuing struggles over resources and claims to authority. Within and between sectors there are organisations that actively seek to 'play the Europe card' by orienting themselves strongly to the flow of EU initiatives and attempting to exploit to the full the emerging opportunities presented by the process of integration. Equally, there are organisations that passively or actively resist such pressures, play down or ignore the relevance of EU initiatives and stress the continued significance of traditional, domestic practices and procedures. These opposing tendencies produce, in aggregate, a pattern of 'lead' and 'lag' organisations and sectors at the domestic level. (The conceptual basis for our methodology is detailed in Buller, Lowe and Flynn 1993.)

This was the perspective adopted in analysing the results from the various sectoral studies. Comparative organisational analysis was pursued along two dimensions:

- *Inter-sectoral comparisons.* Europeanisation is differentially impacting on different organisational sectors and altering the relationship between them. Comparison between the sectors (local authorities, statutory agencies, interest groups) helps to identify changes to the structure and style of the environmental policy field as a whole within Britain.

- *Intra-sectoral comparisons.* The differential response of organisations of
  the same type to the opportunities and challenges posed by European
  integration has similar consequences for intra-sectoral relations. (Some
  environmental groups, for example, have become much more active in
  European political processes than others, with significant consequences
  for the structure of the British environmental lobby.)

THE OPPORTUNITIES OF THE EU

The European Union and its developing environmental policy competence
have provided a new sphere of political opportunities for organisations in
the environmental field. They have responded to these expanding opportunities according to their own particular circumstances, and it is possible to
identify both push and pull factors.

The Conservative administrations of the early and mid-1980s – with their
ideological commitments to deregulation and rolling back the state, their
emphasis on industrial competitiveness and their rejection of aspects of the
traditional consultative style of British government – presented an often
unreceptive edifice to environmental concerns. McCormick argues that
'because the British government has made few concessions on domestic
environmental issues, British environmental groups have increasingly seen
Brussels as a court of redress and a means of out manoeuvring the government' (1991: 132–3). Whereas business and industrial groups have tended to
view central government Departments and Ministers as defenders of their
interests in Europe (Chapter 9), other organisations in the environmental field
have rarely trusted the government to push a pro-environmental agenda. On
the contrary, environmentalists coined the phrase 'Britain, the Dirty Man of
Europe', to challenge the complacency of British officials and politicians over
domestic arrangements for environmental protection (Rose 1990). As Britain
acquired a reputation as a lag state in the development of EU environmental
policy, so British organisations in the environmental field – unlike their
counterparts in, say, Germany, Denmark, or the Netherlands, but in keeping
with those from other lag states – generally came to see the EU as a positive
force for raising domestic environmental standards and sought to use the
institutions of the EU to this end (Sands 1990; Rucht 1993).

As a result, British organisations tend to have a high profile within European networks. They have always been prominent in the European
Environment Bureau (EEB), the oldest European environmental network,
and recently in the formation of new networks such as BirdLife International and Transport and the Environment (see Chapter 6). Likewise British local authorities have set up the largest number of regional offices in
Brussels of any country (John 1994) and are the largest number of participants in European urban environmental networks (Ward and Williams

1997). Similarly, British agencies have been to the fore in establishing and developing the IMPEL Network – the network of official national agencies and inspectorates set up to coordinate the implementation of EU environmental regulations (see p. 99).

The apparent ease of British organisations in dealing with European institutions has been explained in terms of similarities in the political style of interest mediation (Hey and Brendle 1992). Both systems have strong executives, and influence and negotiation are controlled administratively. Therefore, UK environmental groups are quickly familiar with the EU system and have an initial advantage over other national organisations (Chapter 6).

Another push factor is that access to government in the British system is largely discretionary. There is not the right to challenge government decisions that is provided elsewhere through such constitutional mechanisms as formal representation on legislative committees, administrative courts, referenda, federal procedures and the like. Thus, for interests that lack a significant sanction, there is often little that can be done to press their case further once the relevant senior civil servant or Minister has reached a decision on a particular matter. What European institutions provide for British organisations, therefore, is not just additional channels of influence, but a court of appeal beyond the British government. This has both diminished their dependency on discretionary access and provided them with opportunities to challenge domestic policy. One consequence is that, in some fields, domestic policy communities which were previously closed to environmental interests have been prised open or destabilised within a European framework. (The water policy community seems to be a classic example: Ward, Buller and Lowe 1995; Richardson and Maloney 1994.)

The style of European legislation – with its tendency to detail in legal form the specific standards to be achieved – has facilitated this process and has enabled organisations in the environmental field to adopt new tactics and new issues. The tradition in British legislation had been to leave considerable discretion to Ministers and officials. This fostered a style of political management characterised by internal administrative arrangements within relatively closed communities. In many cases for the first time, EU environmental Directives now furnish formal yardsticks against which outsiders can judge progress and performance. The regular publication of information collected to demonstrate to the Commission compliance with standards has meant that previously submerged issues have been opened up as foci for oppositional politics and debate. Directive 90/313 on Freedom of Access to Environmental Information and the Environmental Impact Assessment Directive ensure the public availability of relevant information. Hence Europeanisation has fostered a watchdog approach among environmental interests, giving them much greater leverage in the implementation of policy.

PHILIP LOWE AND STEPHEN WARD

THE DIFFICULTIES OF OPERATING IN THE EUROPEAN ARENA

While European integration opens up new political opportunities, engagement in European fora incurs costs that are not trivial, and significant obstacles. The main difficulties can be summarised under the three headings of resource problems, structural problems, and the complexity of EU policy processes.

A European strategy involving active lobbying of EU institutions and transnational networking is costly in both time and money. Key costs include the need to acquire up-to-date information and intelligence, and to employ staff with European experience, expertise and language skills. For many organisations the resource requirements of a European campaign are either beyond their means or do not seem justifiable. When crucial resource decisions are being made, European activities can easily be viewed as a peripheral luxury or appear distant from the day-to-day concerns.

Sustained engagement in European fora involves not only resources but also the establishment and maintenance of structures of representation to support political access to EU institutions and across the Member States. The European Commission's preference is to deal with EU-wide federations and any effective European lobbying is likely to involve transnational coalitions. Many studies, though, have illustrated the incipient weaknesses of European lobbying federations, including tendencies towards: chronic under-resourcing; lowest common denominator positions; a reputation for ineffectiveness; and an inability to contain maverick activity by members. Mazey and Richardson (1992b) have argued that one of the comparative strengths of environmental organisations, in the European context, is their skill at coalition building, enhanced by the transnational character of environmental issues and the relatively low level of competitiveness between organisations. However, Hey and Brendle found European umbrella organisations in the environmental field to be incipiently weak:

> Umbrella organisations are faced with conflicting goals and difficulties that are partly the result of their structure . . . They must mediate between very heterogeneous interests comprising a very large number of members. The small organisations need the service functions of an umbrella organisation the most, but the umbrella organisation needs the support of the larger, financially sound organisations. The larger organisations have less need of the umbrella organisations, and the smaller organisations cannot provide the umbrella organisations with the necessary political weight alone.
>
> (1992: 13)

In many respects, these difficulties simply reproduce at a transnational level the problems experienced by national umbrella organisations (Lowe and Goyder 1983) but with the added problems of having to overcome differences in the participants' cultural and national backgrounds.

Even if organisations have the resources and structure through which to move into European fora, the European policy making process can seem daunting. While the basic rules of lobbying may share some similarities with those of the domestic arena, the European polity is considerably more complex (van Schendelen 1993; Chapter 3). The relative openness of the EU's institutions and the multiple access points pose novel problems. The policy process involves a range of institutions with very different lobbying climates and encompassing elements of fifteen separate national political systems and cultures. Compounding this complexity is the highly sectoralised nature of the Commission and the Council of Ministers. Thus policy is often fragmented into specialist fields of competence each with considerable independence, and there often appears little co-ordination between them. This presents particular difficulties for environmental concerns which tend to cut across bureaucratic and sectoral boundaries. It often means that a number of Directorates-General may have an interest in an issue; issues may be pursued in parallel by different DGs; and there may be tussles over competence (Butt Philip and Porter 1993). Effective lobbying may require maintaining multiple access points. However, European policy making fora are increasingly crowded by the growing number of lobbyists, which has created considerable competition for issue space.

These problems of resources, structure and European policy making mean that the national (governmental) route into Europe is still of considerable importance, despite the increase in European lobbying. Organisations have their established channels of access in the domestic arena and they naturally find it easier to use these than to develop new ones at the European level. This persistence in lobbying through official national channels is reinforced by the continuing centrality of member states in the decision making structures of the EU (Grant 1993). Furthermore, it is member states that retain control over the implementation and monitoring of European legislation. In consequence, the establishment of new points of access at the European level must go hand-in-hand with the maintenance of existing national channels. European lobbying can be no substitute for, but must be additional to, national lobbying.

SURVEY RESULTS: OVERVIEW

Among all the organisations that we surveyed there was a consciousness of both opportunities and difficulties created by European integration and of the specific challenges posed by the particular balance between these for individual organisations. Many organisations are in the paradoxical position of considering the European Commission more environmentally concerned than the UK Government; of regarding the European Union as a more influential force in environmental policy than national government; but of finding access easier to Whitehall than to Brussels (see Table 5.1).

The working out of this paradox for each organisation leads to a strongly marked pattern of lead and lag organisations within sectors, the key mediating factor being resources. By and large, it is the better resourced organisations that are the most active at the European level.

Between the sectors, a second differentiating factor is apparent – namely structural or constitutional constraints. This is the overriding factor for the statutory environmental agencies. It is very apparent that their dependence on their parent Departments for access to government has severely constrained their direct involvement in EU policy making. Environmental groups are least constrained in this regard (although internal structures, particularly of a federal nature, may limit their scope for European involvement) (Ward, Talbot and Lowe 1995).

Across and between sectors we can identify a spectrum of organisational responses to European integration, ranging from minimal receptiveness to active involvement in EU policy processes:

(a) tracking events/gathering information;
(b) involvement in/monitoring of domestic compliance with EU legislation;
(c) domestic lobbying on European matters;
(d) seeking European funds;
(e) European networking;
(f) European lobbying (*ad hoc*);
(g) creating own supranational structures;
(h) sustained European lobbying and agenda setting over wide range of issues.

Table 5.1 Comparisons of European institutions and the UK Government in environmental policy

	Proportion of Yes/No responses (%)	
	Local authorities	Environmental groups
European Commission is more environmentally concerned than UK Government	80:14	60:22
European Union is more influential in environmental policy than UK Government	23:47	49:32
European Commission is more accessible than UK Government	——	10:53

Based upon where their responses lie on this spectrum, environmental organisations can be categorised as follows:

- *The untouched.* These are the organisations which are un-Europeanised and remain untouched by any of the functions (a)–(h). They have no contact with European institutions or policies, which they see as largely irrelevant to their own agenda.
- *Administrative reactive.* These are involved with (a) and (b) in European terms. While they may show varying and fluctuating degrees of awareness of European politics, this does not strongly influence their activities.
- *Watchdogs.* Organisations involved additionally in (c) and (d), while still largely reactive to processes of European integration, find their agendas heavily influenced by European information, legislation and funding, although their activities are largely confined to the domestic arena. Essentially these organisations are involved in implementation and compliance politics.
- *Limited proactive.* Organisations here have moved beyond the domestic arena and into European lobbying encompassing (e) and (f). Their European activities are limited, however, in terms of both issues and range of contacts. Involvement is likely to be *ad hoc*, perhaps centring on particular pieces of legislation, and links into European institutions are neither extensive nor strong.
- *Transnational proactive.* A small elite of organisations are fully integrated and involved in all the functions from (a) to (h). Such organisations twin-track their lobbying domestically and in European fora. They lobby on a portfolio of issues and are capable of dealing with wider political questions of European integration. They have their own supranational structures of representation.

Table 5.2 categorises the European orientation of organisations in the different sectors on this basis. Being the most structurally constrained but also well resourced, it is not surprising that the statutory agencies

Table 5.2 The European orientation of environmental groups

Classification	Environmental groups	Local authorities	Environmental agencies
The untouched	28%	——	——
Administrative reactive	20%	50%	——
Watchdogs	17%	42%	60%
Limited proactive	21%	8%	40%
Transnational groups	14%	——	——

show the least variation between themselves in their responses, falling into the category of either 'Watchdogs' or 'Limited proactive'. In contrast, the environmental groups, with great variability in their resources, show the greatest variation in their responses to European integration, ranging from those seemingly 'Untouched' by it to those that are 'Transnational' in their operations.

The question arises of whether the spectrum of activities (a)–(h) represents a ladder or a series of hurdles. The implication of a ladder of integration is lead and lag organisations, but with opportunities for the latter to follow in the former's footsteps. However, our results suggest increasing differentiation, with little evidence of movement taking place between the categories.

INTRA-SECTORAL ANALYSIS

Local authorities

In the UK, local government has traditionally had a strong role in implementing environmental policy and over the course of the past decade many local authorities have championed progressive environmental policies. At the same time the increasing significance of the EU in the environmental policy field, particularly since the Single European Act of 1987, has considerable implications for local government. It has been argued by some that the EU has acted as a centralising force in environmental policy relations, indirectly diminishing the statutory responsibilities of local authorities. Yet it is possible in recent years to detect the beginnings of a direct relationship between the EU and local government in the environmental sector, which others claim presents authorities with potential opportunities to develop a stronger role (see Chapter 8).

Thus, on the one hand, it has been suggested that the development of policy at the European level reduces the scope for local authorities to act independently and erodes their responsibilities, as statutory powers move upwards. Potentially it also further distances local authorities from the locus of central decision making, thereby threatening to marginalise them from key policy processes. Nigel Haigh has argued that:

> the European Community's environmental policy has certainly involved moving powers upwards from the nation state to a Community level but not downwards from the level of nation state to lower levels. On the contrary, and this is a surprise to many, the hold of the national governments on local and other authorities has increased.

> (Haigh 1986: 205)

On the other hand, the Environment Directorate in Brussels, at least when Carlo Ripa de Meana was the Commissioner, was looking to build constituencies for the development of a more ambitious EU environmental policy. A crucial element was the publication in 1990 of the Commission's Green Paper on the Urban Environment (CEC 1990c) in which the Environment Directorate made a conscious bid to move into the field of urban planning and to make links with local authorities across Europe. This thinking was taken forward in the EU's Fifth Environmental Action Programme, *Towards Sustainability*, to guide EU policy for the rest of the decade (CEC 1992b). In this Programme the Commission hoped to redirect the debate around the issue of subsidiarity to what it described as the 'principle of shared responsibility'. That is to say, the central concern should not be the separation of competencies but rather the appropriate degree and form of co-operation between the Union, member states, local authorities, enterprises and citizens' groups in resolving particular environmental problems. This concept of 'administrative subsidiarity' is obviously attractive to local authorities. Indeed, it has been emphasised that 'local authorities are responsible for implementing 40 per cent of the Fifth Environmental Action Programme' (Local Government Management Board 1993: 28).

The actual picture of local authority involvement in European environmental matters is very patchy. A small number of authorities (fewer than 10 per cent) have a well developed understanding of the implications of EU environmental developments and are able to respond proactively. They are involved in European networks or EU consultative fora. They are largely upper tier or urban authorities, and include Bath, Bristol, Cardiff, Glasgow, Grampian, Gwent, Hertfordshire, Humberside, Kent, Lancashire, Leicester, Newham, and Strathclyde. A considerably larger group of authorities are geared up to deal with the EU largely as a source of funding, including for environmental schemes. Typically they employ a European officer. They can also react to legislation from the EU, but have yet to develop the ability to input on a regularised basis to EU environmental policies. This group includes most of the rest of the upper tier and metropolitan authorities as well as some of the more active districts, particularly in Scotland and Wales. There are still, though, many authorities – about half of the total number – who are marginal to environmental developments in the EU. They feel poorly informed and indeed may well not recognise the European origins of much environmental policy and legislation. Most of these are lower tier authorities, particularly those covering rural districts.

In explaining variations in the level of European environmental activity among authorities, four factors are important:

- *Entrepreneurs and spillover.* Policy entrepreneurs who are active, knowledgeable and have links with Europe concerning EU environmental policy transfer their knowledge to others within the authority, creating an internal European network. These actors are normally senior officers. Where authorities have a general European strategy and links to Europe in other policy areas, this may spill over from one policy arena to another.

- *Funding bids and policy input.*   Those local authorities which have extensively pursued EU environmental funding have gradually expanded this interest into policy matters. The bidding process itself increases awareness of EU policy, since authorities generally develop a background knowledge of the policy area in which they are bidding for funds. Requirements that bids for funds should include international partners encourage the development of transnational networks.
- *Strong environmental orientation.*   Authorities which have been the most environmentally active domestically over a long period have embraced the European dimension relatively quickly and are using it proactively to promote the image of a forward looking, professionally oriented authority. Frustration with central government on environmental issues has led to an increasing interest among these and other authorities in bypassing the centre to deal directly with Brussels.
- *Resources.*   The authorities that are the most proactive tend to be those with greater resources and administrative capacity, as well as the statutory strategic functions that make the Union appear more relevant to the authorities' needs. They are mainly upper tier authorities, metropolitan councils and some of the larger urban district councils.

Statutory environmental agencies

Statutory agencies have traditionally played a distinctive role in the devolved management and administration of the environment in the UK. They have combined a broad range of functions including specialist policy advice to government, and regulatory and management responsibilities in implementing policy, as well as a more general advocacy and promotional role encouraging environmental activity and awareness. Over the past decade, however, European integration has presented challenges to such agencies and their style of operation.

Increased EU involvement in the environmental sector has generally posed two different types of challenge for the agencies. First, new information flows and responses to European initiatives require organisations to make internal administrative and organisational adaptations to deal with them. Second, the increased importance of the EU in environmental policy making requires the agencies to respond to new policy arenas and supranational policy making.

All the agencies have designated an officer responsible for European matters, though for the most part this forms only a part of a wider international role. Normally such officers act as information conduits or as facilitators, assisting colleagues to develop their own European links. Generally they are not themselves responsible for developing European policies for their agency. None of the agencies currently possesses its own office in Brussels, but some have seconded staff there. All the agencies remain to varying degrees dependent on their sponsoring Departments to keep them informed on EU legislative proposals in their particular area. Some of the agencies noted that their reliance on central government was at times unsatisfactory, since lines of communication tended to be rather *ad hoc*, partial and slow. Agency officials thought that it was not simply a case of central government deliberately acting as a gatekeeper with European information, but also of departmental officials not necessarily knowing what is going on in the Commission. This was linked to a reactive tendency to wait for proposals to come down from the Commission. Consequently, all agencies, while accepting government as a source of information, express the need to continue developing their own information gathering activities. A dependency on central government sources of intelligence is seen to put them at a disadvantage (Ward, Talbot and Lowe 1995).

The environmental agencies have, in rhetorical terms at least, noted the importance of the EU in environmental policy and recognised the need to make their ideas known at the European level. Yet, their European contacts tend to be narrowly focused on the Commission, in particular DGXI (Environment Directorate), although some contacts have also been made with other DGs including VI (Agriculture), XII (Science, R&D), and XVII (Energy). The amount of contact is variable. Some agencies have little direct contact; others have fitful, *ad hoc* links. Agencies rarely initiate contact or attempt directly to sell policy ideas. They are seldom involved in the primary agenda-setting stages of European legislation. Officials in the Commission consider the agencies to be marginal to the European policy process. One senior DGXI official summarised a common view: 'We don't make much use of [the UK environmental agencies] or have much contact with them . . . they are not pushing a policy role' (interview, 25 October 1994). Agencies, therefore, often find themselves reacting when formal proposals have been drafted but at this stage they are expected to channel their responses through central government Departments and tend to be excluded from the formal negotiating machinery in Brussels. As officials of the former National Rivers Authority (now the Environment Agency) explained, they brief civil servants from the DoE extensively, but are not involved directly in the specialist Adaptation Committees of the EU. If contact with the Commission is patchy, then involvement with other institutions of the EU is even more spasmodic. While the agencies acknowledge the developing importance of the European Parliament and express a wish to build closer links, very little is being done practically. They are held back in part through a concern not to become embroiled in partisan politics.

There are significant variations between the UK agencies in their responsiveness to and involvement with the EU. These differences in approach revolve around the interpretation of opportunities that the EU presents, but equally reflect the ability of individual agencies to overcome the difficulties that European integration poses for their domestic position, in particular the strains it places on relations between them and UK government Departments. The main factors which shape agency–EU relations and explain some of the differences between the agencies are:

- *Remits and responsibilities.* With UK legislation increasingly originating from Brussels, the more adventurous agencies interpret this as giving them latitude to involve themselves in European debates and see this as a logical and quite proper extension of their domestic role. More cautious agencies only become involved in European matters on a reactive basis, as and when requested to do so by their sponsoring Department. The legal framework has been most extensively Europeanised in the pollution field, followed by the nature conservation field, with the heritage field least affected. The agencies therefore have differing degrees of direct European responsibilities as the competent authority for particular Directives.

- *Relations with parent Departments.* All the agencies see it as important to maintain a good relationship with their parent Department by keeping it regularly informed of what they are doing with regard to the EU. There are variations between them, though, over whether policy stances can or should be taken in advance of the government's and sold directly to the Commission. The most reactive agencies take a strictly formal line that any policy objectives with regard to Europe should be channelled through the Department. Other agencies believe that they can twin-track their ideas through government and via other routes to influence the EU. Agencies also referred to perceived differences in attitudes of the various sponsoring Departments. The DoE is regarded generally as being cautious and mildly suspicious of European activities of agencies within its field, whereas the Scottish and Welsh Offices are thought to be more relaxed, and are keen to see their agencies helping promote a regional presence in Europe. A sponsoring Department is also likely to take a more relaxed attitude where agencies are being proactive in another Department's area of responsibility. Consequently, the DoE appears less cautious when its agencies are active in other fields such as the European agricultural debate (see Chapter 2).

- *Peculiarly British structures.* The range and type of agencies that exist in Britain have few counterparts within Continental Europe. It is widely agreed that the intricacies of the British system and the distinction between agencies and government departments are not well understood in European circles. The confusion is compounded by the varied functions and geographical responsibilities of the agencies, which make it difficult for them to speak on an authoritative, UK-wide basis.

- *Lack of involvement in European networks.* An aspiration of the European Commission is to deal with expert, representative European networks. Likewise, effective action at the European level depends on building transnational coalitions. However, the agencies have found networking difficult, partly because of the lack of analogues in other European countries, but also because they feel uncomfortable about forming alliances which may seek to lobby not only the Commission but also other member states' governments. A potentially significant development, therefore, is the European Community Network for the Implementation and Enforcement of Environmental Law, the so-called IMPEL Network, which involves the Environment Agency from the UK. The network originated in 1992 largely from concerns over compliance problems with pollution regulations (see Chapter 2). Its subsequent slow development is symptomatic of the problems of obtaining workable compliance and implementation structures on a transnational basis. In 1993, efforts were made to bring together environmental advisory bodies from the member states in what has become known as the European Environmental Advisory Councils. It has held a number of conferences to further co-operation and in 1996 it was decided to seek funding from the European Commission to establish a common Secretariat. From the UK, English Nature has played a leading role in these developments.
- *Continual administrative revolution.* Agencies' abilities to deal with European issues are not assisted by the almost continual cycle of agency reform initiated by government. Officials from NRA and HMIP believe that the new environmental agency which was set up in 1996 would eventually strengthen their contacts with the EU. The creation of an integrated pollution agency, with a large degree of technical expertise under one roof, should make it more difficult for the DoE to maintain that policy development abides solely within the Department. In the short term, however, the experiences of the nature conservation agencies following reorganisation in the early 1990s are not encouraging, suggesting that structural and administrative upheaval may disrupt strategic thinking about Europe, by making agencies too introspective and preoccupied with internal administrative and domestic matters.

Environmental interest groups

The European Union and its developing environmental policy competence have provided a new sphere of political opportunities for environmental groups. There is a growing national but also comparative literature which analyses the ways groups have responded in terms of the relative possibilities open to them in the opportunity structures of national politics (Rucht 1993; Mazey and Richardson 1992a; Dalton 1992). That literature largely draws upon the experience of a handful of elite or vanguard groups. Not only does this convey a very partial perspective on the experience of the environmental movement overall, it also does not help in understanding the obstacles to European participation or the effect of European integration on the internal structural differentiation of the environmental lobby. Our survey therefore attempted to gain a broad perspective on the effects of European integration on the environmental lobby, including the diversity of reactions and strategies between different groups, to move beyond partial analyses and prognostications based on selective examples.

An initial, striking result of the survey was the varied involvement it revealed. Some 47 per cent of the groups questioned had no contact with European institutions. Only a small elite of groups (14 per cent) appear well integrated with consistent, wide ranging connections into European fora. This rather patchy involvement does not reflect a lack of appreciation of the role of the EU in general or of the Commission in particular in environmental policy and legislation. Overall environmental groups now rate the EU as a more influential policy actor than the UK Government in the environmental sector; and a clear majority judge the Commission as more environmentally sympathetic than national government (see Table 5.1, above). Even groups with little or no contact with European institutions may need to respond to policies emanating from those institutions. Some 28 per cent of the groups claimed to have no dealings with European issues. The rest have to manage an often growing workload stemming from European legislation or agendas. To do so, some have made structural adaptations: internally, by developing mechanisms of managing European issues and information flows; and/or externally, by networking and coalition building with other organisations to gather information and lobby on a collective basis.

Internally, the majority of groups handle European issues on an *ad hoc* basis, with no overall European policy co-ordination: in the larger groups, issues are generally processed by the relevant policy specialist; and in the smaller groups, by a senior official. Only 18 per cent of groups had a designated European officer co-ordinating European policy information and responses within the organisation, and these were mainly the very large elite groups such as RSPB.

Externally, British environmental groups display a high level of networking. Two-thirds of the groups surveyed claimed membership of a European network and altogether twenty-one different networks were recorded. Of these, a third of the groups belonged to the European Environmental Bureau (EEB), the largest European network. Though the extent of networking appears impressive, it masks differences in the objectives of the networks and their actual involvement in European policy making. The role of the majority of the networks centred on information exchanges rather than lobbying or representation in Brussels. Only seven networks were described by members as having a lobbying role: Coordination Européenne des Amis de la Terre (CEAT)/FoE Europe; Greenpeace EU Unit; BirdLife International; European Environmental Bureau (EEB); Transport and Environment; FACE[4]; European Sustainable Agriculture Group. The case of EEB is particularly interesting, since although it has a lobbying role the overwhelming majority of its UK members do not see it in these terms but as an information vehicle. The larger groups, now with their own networks, use it as an additional source of information and coalition building, whereas the smaller groups, who have no direct involvement in Europe use it mainly to keep in touch with EU policy.

The survey underlined the continuing importance of the European Commission and within it, of DGXI, to environmental groups, as the primary focus of lobbying efforts. Over half the groups concentrated on the Commission in their European lobbying: of these 98 per cent had links to DGXI. The next most sought-after Directorate was Agriculture (DGVI) with which 38 per cent of the groups with links with the Commission had contact. The limited range of contacts within the Commission is revealed by the fact that only 16 per cent of groups had regularised contact with more than two Directorates. WWF and RSPB stand out in having extended their range of contacts into other Directorates, notably DGXVI (Regional Policy) and DGVII (Transport). The other main economic Directorates remain largely untouched by British environmental groups. It would appear that most environmental groups are still heavily dependent on DGXI for access, and that the environmental lobby is still ghettoised within the Commission.

Environmental groups have extended their range of lobbying to other institutions, in particular the European Parliament. Some 39 per cent of all the groups surveyed – that is, 71 per cent of those with some European involvement – have had experience of lobbying the Parliament, which is perceived as being the most open and environmentally attuned of all EU institutions (Judge and Earnshaw 1993). Its Environment Committee is widely acknowledged to be one of the most active and has a strong British presence. However, although the Parliament's role was strengthened by the Maastricht Treaty, it is still regarded primarily as an arena for raising issue publicity and gathering information, but also increasingly for pressurising the Commission.

In explaining variations in the level of European activity among environmental groups five factors are important:

- *Entrepreneurs and spillover.* Involvement in European institutions is often developed, in the first instance, through individual contacts and personal initiative. A key role is thus played by entrepreneurial senior officials within a group who make the link between European and domestic issues. Often this starts with a specific piece of EU legislation whose significance draws the group into the European arena, helping them thereby to recognise more generally the importance and relevance of European institutions to their wider agenda.

- *Funding attractions.* An important factor in attracting some groups is the possibility of European funding. Some 41 per cent of the groups surveyed had made applications for European funds and just over half (22 per cent) had been successful. These were largely the groups who were proactive in European policy. Significantly (and in contrast to local authorities), groups do not in the main pursue EU funding as an end in itself, separate from policy inputs; instead they tend to see EU funding as a means to influence policy, either directly (through conducting a piece of applied research or monitoring for the Commission) or indirectly (through providing the resources, contacts and access for European lobbying).

- *Resources.* The groups that are most active in Europe tend to be the biggest ones, the wealthiest ones, the ones with the largest staff, and the ones with their own European networks.

- *Remits.* Some groups see less significance in the EU than do others. Among the least active in European fora are those with a focus on built environment/heritage concerns, a field not so strongly developed by the EU. Others apparently little touched by Europeanisation include groups with a parochial agenda who are well embedded in Britain's local planning system, such as the Civic Trust. In contrast, the more active groups in Europe tend to be those with a more scientific or international orientation, mainly concerned with wildlife or ecological issues.

- *Structures.* European involvement is more straightforward for large, centralised organisations (such as the RSPB, WWF, Greenpeace) than for groups with (semi) federal structures (such as the Wildlife Trusts, CPRE, the Civic Trust, Council for National Parks). The former are able to generate the necessary resources and have more flexibility in deploying their staff in European fora. The latter tend to be hampered by their federal structures in making such commitments and developing trans-European structures.

CONCLUSIONS: STRUCTURAL CHANGE IN THE ENVIRONMENTAL POLICY FIELD

The structural consequences are clearest within the environmental groups sector. Most groups lack the capability or the desire to move much beyond (c) (domestic lobbying) on the integration ladder (see p. 92), while a small number of groups are fully integrated at the European level. The effect of Europeanisation has essentially been to strengthen the position of groups that already were in the elite domestically. It has done this by furthering the process of institutionalisation and professionalisation within the environmental lobby. There is not much room for voluntarism, committed amateurs or grassroots involvement when groups are attempting to operate at the European level. The costs of being involved also effectively pre-select those professionalised groups with a considerable capacity to raise and deploy their own resources. In turn, though, this involvement entrenches the dominant position of a handful of groups (FoE, WWF, Greenpeace, RSPB, CPRE and Transport 2000) that are highly active in both national and European policy processes.

The structural differentiation of the environmental lobby under the impact of European integration is producing new divisions of labour. In general, environmental groups are more dependent upon one another than they are on government for access and information on Europe. The larger groups that are active transnationally act as gatekeepers to European information for the smaller groups. In return, they rely on the smaller groups who are restricted to domestic politics for specialist niche information, for co-operation in broad domestic lobbying coalitions and for fulfilling national and local watchdog roles. Increasingly, many of the smaller groups are concentrating on implementation politics. Within Brussels, these developments are undermining the viability of the European Environmental Bureau as the broad umbrella body for the environmental lobby and leading to its displacement as the peak organisation by a number of independent (but co-operating) specialised supranational networks set up by the elite groups. This move from an inclusive to exclusive structures can be seen as a form of closure, because of the prohibitive costs to most groups of establishing their own supranational structures of representation. It is paralleled in other interest group fields which have likewise seen the appearance of more 'European' structures (Andersen and Eliassen 1991; Butt Philip and Porter 1993; Greenwood *et al.* 1992; McLaughlin *et al.* 1993).

Although we have concentrated on the environmental groups sector, structural change is also evident within the other sectors we have studied and between them. For example, European integration has tended to divorce the policy advice and promotional roles of environmental agencies from their regulatory and executive functions. The latter functions have increased as a result of European legislation since the agencies are seen to offer to government an efficient means of centralising policy implementation. However, their policy advice and agenda-setting roles have shrunk, as the focus of policy making has shifted to Brussels and they have been marginalised from the policy process. Their traditional role and influence in British politics depended on the way in which they combined advisory and promotional functions with regulatory and management functions which meant that they accumulated considerable policy expertise. The split in these functions under the impact of European integration is helping to set the seal on the declining influence of the agencies in UK environmental politics.

What is perhaps most striking is the change in the relative influence of the agencies and the elite environmental groups (see Chapter 12). In the past, groups were often dependent on the agencies for access to government. Now the elite groups have forged their own channels of access to European institutions, where their presence, range of contacts and influence greatly exceed that of the agencies. In the words of one Commission official: 'the voluntary organisations are streets ahead' (interview, 26 October 1994). As the groups have emerged with a significant agenda-setting role at the European level, this has reinforced their dominant position in this regard in domestic politics, further eclipsing the agencies.

NOTES

1   A questionnaire was sent to sixty-two national environmental groups. The list was drawn up from *Who's Who in the Environment*. It covered the active voluntary organisations with a UK or English, Welsh or Scottish remit. The response rate was 75 per cent. The survey was complemented with detailed interviews with eleven groups (British Association for Shooting and Conservation, CPRE, FoE, Greenpeace, Marine Conservation Society, National Trust, Pesticides Trust, RSPB, Surfers Against Sewage, Transport 2000 and Wildlife Link).

2   A questionnaire was sent to all local authorities in England, Wales and Scotland (516 in total). Since both environmental and European issues cross departmental boundaries, the questionnaire was aimed at corporate environmental officers, who were likely to have a knowledge of environmental affairs not restricted to a single department. In total 278 completed questionnaires were returned representing a very respectable response rate of 54 per cent. Interviews were also carried out with the relevant European and environment policy officers in the local government associations (Association of County Councils, Association of District Councils, Association of Metropolitan Authorities and the Convention of Scottish Local Authorities), the Local Government International Bureau (LGIB) and the Local Government Management Board. These interviews were used to gain an overview of relations between local government and the EU in the environmental field. A number of interviews were also carried out with local authority representatives on EU consultative groups to establish an inside and elite view of relations with the Community. In total around twenty interviews were conducted.

3   An initial questionnaire survey was undertaken of twenty-two national and country (i.e. English, Scottish and Welsh) agencies in the countryside and environmental fields. Most of these were independent statutory bodies but we threw the net wider to include departmental bodies with equivalent functions, such as the Forestry Commission and Her Majesty's Inspectorate of Pollution. There was a response rate of 86 per cent. In addition, in-depth interviews were conducted with key staff in the following agencies: National Rivers Authority, Her Majesty's Inspectorate of Pollution, the Health and Safety Executive, English Heritage, English Nature, Forestry Commission, Countryside Commission, Scottish Natural Heritage, Countryside Council for Wales and the Joint Nature Conservation Committee.

4   FACE is the Federation of Hunters' Associations in the EC. It is the European network to which the British Association for Shooting and Conservation belongs.

6

THE ENVIRONMENTAL LOBBY

Tony Long

This chapter covers the development of the environmental lobby at the European level. It describes the thickening web of relations between environmental organisations working in Brussels over a twenty-year period and attempts to show how far environmental groups from the United Kingdom have been influential in this process. The factors leading to the decision by one such organisation in the UK – the World Wide Fund for Nature (WWF) – to establish a presence in Brussels are examined. The overall conclusion is that, although the environmental movement has responded to the challenge of co-operating and building a European network, only limited amounts of the resources and skills available at the national level are brought to bear on the European Union policy process.

ORIGINS AND DEVELOPMENT OF THE EUROPEAN ENVIRONMENT LOBBY

For almost fifteen years, the European Environmental Bureau (EEB) with its European network of member groups was the only significant environmental organisation in Brussels (Lowe and Goyder 1983). The origins of the EEB can be found in an initiative by Canadian and US environmental groups in inviting their European colleagues to a meeting in Brighton, UK, in January 1973. Issues under discussion at the meeting included nuclear (energy) problems, ocean pollution, international conventions, lobbying of international bodies and environmental impact assessment (Vonkeman 1994).

Almost two years later, in December 1974, the EEB was formally launched with a membership of twenty-five environmental organisations.[1] The inaugural statement was brief but its language was extremely prescient. It stated:

European environmental protection associations, working jointly with other organisations active in the social field, have a role to play in promoting a new future for Europe . . . [they] have decided to work together with the following objectives:

- to promote a lifestyle which is sustainable in the long-term and equitable;
- to promote protection and conservation of the environment, restoration and better use of human and natural resources, particularly in the context of the policies of the EEC;
- to promote, launch and coordinate research and studies to reach these objectives;
- to make all necessary information available to its members and other organisations able to contribute to the realisation of the objectives of the Bureau;
- to use any appropriate educational or other means to increase public awareness about these issues;
- to draw up and transmit to the appropriate authorities, in particular in the EEC context, recommendations aimed at pursuing the Bureau's objectives.

(first EEB statement, 18 December 1974)

In a joint article in the EEB's twentieth anniversary publication, the three successive Secretary Generals describe three phases in the EEB's evolution (David et al. 1994).[2] The first phase was getting the environment accepted as one of the responsibilities of the Community, and getting environmental groups to accept Europe as a focus of attention. In the second phase the challenge was to change environmental groups, which tended to be rather introverted, into a broad social movement. In the most recent phase, the challenge has been to deal with a widening agenda and the broadening of the Union into a major international institution. The three authors highlight the White Paper on the Single Market in 1985 as being a significant moment in the environmental lobby's attitude towards Europe. There was a realisation then of the scale of the changes being prepared and their potential environmental impact.

The surge in interest in the Single Market in the mid-1980s coincided with the accession of Greece (1981) and of Spain and Portugal (1986) as new member states alongside the passage of the Single European Act in 1986. In this turmoil, changes were taking place too in the environmental movement. The mid-1980s were a time which Hey and Brendle (1992) describe as marking a far-reaching transformation from what they call an environmental 'movement' into an 'institution.' This transformation was a response to three important developments:

- the substantial growth of large environmental organisations which by the early 1990s could boast a membership of more than 10 million people in the European Community;
- the political career of 'environmental issues' which required a shift in lobbying strategies; and
- the internalisation of environmental policy, requiring co-ordinated action on various levels.

It is not surprising that the second half of the 1980s saw an explosion of interest among environmental groups in the policy making and legislative functioning of the European Community. Several groups decided to establish an institutional presence in Brussels then. The first to do so (see Figure 6.1) were the large international organisations – CEAT (Coordination Européenne des Amis de la Terre, better known as Friends of the Earth Europe), Greenpeace, and the World Wide Fund for Nature (WWF). These groups were joined for a time by a German-based organisation with a number of European conservation projects, the Stiftung Europäisches Naturerbe (Foundation for European Conservation Heritage). Climate Action Network, set up in 1989, heralded a shift towards the establishment of specialised lobbying groups.

A second wave of new arrivals occurred between 1991 and 1993. This was marked by the opening of Brussels-based offices by other more specialised networks, including the Transport and Environment Federation and

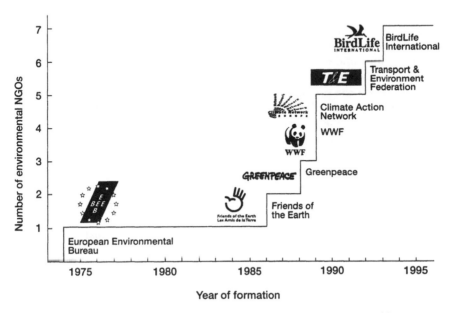

Figure 6.1 The growth of the Brussels-based European environmental lobby

BirdLife International. The European Habitats Forum, a network of mainly international environmental groups concerned with nature conservation, was also created at this time but without having a full-time institutional presence in Brussels. Two smaller groups, the Environment and Development Resource Centre (EDRC) and the European Bureau for Conservation and Development (EBCD), also opened offices in Brussels in this period; they were perceived as having more of a research and advocacy function and lacking some of the more typical features of non-governmental organisations.

There are several reasons to account for the explosion of interest in European-wide campaigns and lobbying by environmental groups at this particular time. The new Title added to the Treaty of Rome by the 1986 Single European Act introduced an explicit environmental power for the first time. Terms such as prevention at source, polluter pays, and precautionary approach were included in the legal basis of the Community. The crucial idea of integrating environmental policy into all the Community's other policies, first given prominence in the Commission's Third Environmental Action Programme, was now given legislative weight in the Treaty. Although expressed in fairly vague terms, the integration requirement in Article 130r put an end once and for all to the idea that environmental policy and legislation could be confined to a discrete box managed by Directorate-General XI. In other words, environmental groups had many more targets for their lobbying in Brussels, and more legislative muscle for their arguments, than had existed prior to the Single European Act.

The amount of EC environmental legislation started to expand enormously from this point (see Chapters 1 and 2). The impacts on the environment of sectors and Community programmes previously largely immune from environmental scrutiny, such as agriculture, energy, transport, regional policy and overseas development, became better understood. There was a clear advantage for environmental groups in tackling these impacts at the level where the problems originated, which often was more likely to be Brussels than the member state (Mazey and Richardson 1992b). In the public debates about the environment in the mid-1980s, there was a growing recognition simply that environmental issues had transfrontier impacts. The acid rain controversy of the mid-1980s pitting Scandinavian countries against Britain was a case in point. As more money began to be available for the environment in EC budget lines, often inserted by a sympathetic European Parliament, some organisations saw an advantage to step up their presence in Brussels for fundraising purposes.

For all these reasons, an exclusive focusing at national level for some of the more policy-oriented environmental groups was no longer adequate. This did not mean, however, that all groups in all countries found the transition from national to transnational thinking and action an easy one to make. The adaptive changes owe much to the political context and tradition of each country.

ADAPTIVE CHANGES AT A NATIONAL LEVEL

The different political cultures operating in the member states play an important role in determining the ease with which environmental groups are able to adapt to the political culture of Brussels (Hey and Brendle 1992) This is because the national political system is a crucial determinant also of the political styles, strengths and weaknesses, main levels of action and the issues selected for action, even if most groups do not question them. Of the four countries studied by Hey and Brendle – the United Kingdom, The Netherlands, Spain and Germany – the authors concluded that the environmental movement in the United Kingdom was the best placed for adapting to Europe.

The authors advance a number of reasons to explain their thesis. First, the UK and Dutch groups are believed to share a 'success-oriented' political style which places a high value on target feasibility and realisation of policy goals. This is in contrast to the more highly principled, 'politically correct' positions of the German and Spanish environmental groups. This feature means that an organisational structure exists in the UK and the Netherlands that is directly geared to influencing policy. The Dutch and British groups, therefore, have a political style and structure that allow them to slot more easily into the policy processes of Brussels.

Hey and Brendle then contrast what they call 'open' and 'closed' political systems at the national level. They conclude that the Dutch and British systems are relatively open compared to those of the other two countries. Environmental groups in the UK have learned to translate their demands and ideas into the 'game rules of the feasible within the given distribution of power', to pursue 'special interest policies' and to raise acceptable individual demands which are negotiated by delving into the logic of the administrative agencies and making them 'palatable' (Lowe and Goyder 1983). The political style of the British environmental organisations is a reaction to a political system with a very strong executive and a weak Parliament. This is in contrast to the Netherlands with its open and integrative political culture, its strong Parliament and relatively weak executive.

Looking at the political culture of the European Community, Hey and Brendle suggest that, with its strong executive and weak Parliament, Brussels is very similar to the British situation. In both cases, political access for the environmental movement is characterised by informal contacts with organisations that have no formal rights or power potential. Environmental policy emerges from a gradual process of adjustment with the executive choosing between what it considers to be either 'constructive' or 'destructive' positions. The authors conclude: 'For British organisations, therefore, the move to Brussels is only a spatial and not a cultural one. This partly explains the familiarity of British organisations with EC institutions' (1992: 6).

Hey and Brendle then look at the importance of political levels to the environmental organisations. They argue that the national level in Britain is the most important policy level. It is here that environmental groups concentrate their efforts. They contend that the lack of interest shown in the environment by the Thatcher Government in the 1980s meant that UK environmental groups saw their greatest possibilities for exercising influence in co-operating with the European Commission. They continue:

> Environmental protection innovations for Great Britain come from Brussels. The British Government's lack of action can be exposed by [the environmental organisations] co-operating and frequently exchanging information with the EC Commission. The British environmental organisations 'use' the EC to improve their position nationally. They have relatively little interest in the EC politics themselves.
>
> (1992: 6)

If Hey and Brendle's overall thesis that UK environmental groups are the best adapted to working in Brussels is correct, this should be borne out by evidence of the success of their lobbying activities at the EU level. The fact is that evidence to support that claim is rather thin. The UK groups may possess particular advantages, and have particular motives for bypassing their national government, to lobby in Brussels (see Chapter 5). In practice, however, British environmentalists are not especially prominent in the corridors of Brussels institutions compared with other nationalities. The explanation is probably two-fold. First, Hey and Brendle may have assumed too simplistically that familiarity with national level politics and campaigning in Britain translates easily into a multicultural and multilingual political context. In practice, British pressure groups face similar handicaps, as do British officials, in dealing with the complexities of European political and cultural traditions (Potter and Lobley 1990; Mazey and Richardson 1992a). Second, unlike their colleagues in mainland Europe with common frontiers, common languages and, in some cases, common geophysical features such as shared river basins, most UK environmental groups have less motivation to join in transnational policy making.

This is not to deny that the UK environmental movement has had some influence on Community policy. Early examples include the role of the Royal Society for the Protection of Birds in initiating the Birds Directive (79/409/EEC) and the influence of the Campaign for Lead Free Air (CLEAR), under the leadership of Des Wilson, on the Lead in Petrol Directive (85/210/EEC). It was UK animal welfare groups which were largely behind the restriction of the import of seal skins into the EC (Directive 85/444/EEC). A notable success in the mid-1980s for UK campaigners was the introduction of so-called environmentally sensitive areas (ESAs) brought in by virtue of the Regulation on improving the efficiency of agricultural structures (797/85). The pressure in the UK to secure recognition of environmental objectives in the Common Agricultural Policy was given added weight at the time by the national controversies surrounding the threats to nationally important landscapes and habitats, such as the Halvergate Marshes in the Norfolk Broads, from agricultural intensification (Baldock *et al.* 1990; Chapters 12 and 13).

A less well known success of the UK environmental movement has been the influence it has exercised in making the Structural Funds more sensitive to environmental concerns. The alarm about the potential for these large funds to harm sensitive habitats, particularly in the Southern member states, was sounded by a joint report of the Institute for European Environmental Policy (IEEP) and WWF looking at the impact of the Integrated Mediterranean Programmes (IMPs), the precursors of the first reform of the Structural Funds in 1988 (Baldock and Long 1987). The eventual strengthening of the environmental protection requirements in the second revision of the Structural Fund regulations in 1993 (Regulation 2081/93), which involved environmental organisations from across Europe, can be traced directly to the early influence of the UK environmental groups, including the important part played by the Royal Society for the Protection of Birds (Long 1995).

The Structural Funds provide a clear example of where domestic interests were less important to the UK environmental groups than the perceived threat to European natural areas. This is contrary to Hey and Brendle's claim that the UK environmental movement was interested in European politics only to the extent that it improved its own national position. The fact is that the early symbols of the destructive effects of the Structural Funds were not UK examples at all but rather the actual damage caused in often extremely remote areas of southern Europe, such as the Prespa National Park in northern Greece close to the border with Albania (Mazey and Richardson 1994). In drawing attention to the environmental effects of these large regional development programmes, UK campaigners were also directly and indirectly questioning the openness and accountability of the Community and member state institutions responsible for their implementation.

Indeed, besides the issues and the campaigns themselves, another way of looking at the European influence of UK environmental groups is to see how far they have sought to develop new, as opposed to using existing, lobbying structures for furthering their objectives. The following sections – a case study of WWF and adaptive changes in Brussels – provide examples of both strategies.

ADAPTIVE CHANGES: A CASE STUDY OF WWF

The following case study focuses on the decision in the latter part of the 1980s by one organisation, WWF, to establish an institutional presence in Brussels. In some senses, WWF is not representative of other UK environmental groups. First, it has a more international outlook than many other groups, for instance through its considerable involvement in overseas development projects. Second, its reputation in the UK is for field-level project work backed up by strong scientific and funding support rather than for policy activities, although such activities are becoming an increasingly important aspect of the organisation's work.

The decision by WWF to open a full-time office in Brussels was taken in 1989. The initiative was supported by the WWF International Secretariat based in Switzerland as well as several of the larger national WWF organisations in Europe. Nevertheless, the initial impetus came from WWF-UK, which also contributed most of the original costs of setting up the office, as well as providing the first two full-time members of staff in 1989 and 1990 respectively.

The impulse behind the decision to open a Brussels office was the perceived importance of the European Community's role in official development assistance, then amounting to about 2.5 billion Ecus per annum flowing largely to what at that time were the sixty-five African-Caribbean-Pacific (ACP) countries. WWF had a dual interest in this expenditure. The first was to try to influence the EC institutions to tighten environmental procedures, for instance through mandatory environmental impact assessments. The second reason was to try to lever some of the considerable sums of EC aid money to be spent on conservation and development projects, some of which it was hoped might be obtained by WWF in support of its own international projects.

The Fourth Lomé Convention then being negotiated between the EC and the ACP countries was seen by WWF as being an ideal target for securing strengthened policy commitments to the environment. This lobbying activity, directed particularly at the Development Directorate (DGVIII) in the European Commission, was designed to build on the earlier successes of the European environmental movement, working closely with its counterparts in the USA, in raising the profile of environmental issues with the World Bank and other regional development banks in the second half of the 1980s. When the Fourth ACP–EEC Convention was eventually signed in Lomé on 15 December 1989, an Environment Title with nine articles had been added to the Convention for the first time, including a provision for mandatory environmental impact assessments for large-scale projects or those 'posing a significant threat to the environment' (CEC 1990b).

Following this achievement, the original rationale for establishing the WWF presence in Brussels, namely influencing overseas development policies, was quickly overtaken by what became to be regarded as more pressing domestic European policy concerns. The WWF Europe/Middle East Programme became operational in 1990 supported by thirteen national WWF organisations in Europe, eight of them within the European Community at that time (the Community then comprised twelve member states).

The aims of the WWF European Programme were threefold: first, to divert conservation expenditure from the relatively rich north and west of Europe to the poorer (but biologically better endowed) south and east of Europe (the so-called 'iron curtain' was being dismantled at that time); second, to focus increased attention and resources on WWF projects of genuine European, rather than national, importance; and third, to encourage cross-border co-operation and projects among the national WWF organisations. Policy activities and a network servicing function, for instance in disseminating news about emerging legislation and funding opportunities, became clear priorities for the WWF office in Brussels (Long 1995).

From being a largely, although not totally, UK creation in 1989, the WWF European Policy Office – as it was later to be called – has now firmly established its European credentials. The two original British staff were joined by other staff from Belgium, Germany, Austria, Spain and the Netherlands in a total full-time complement of eight persons by 1996. Contributions to the operational costs of the Brussels office come from the fifteen national WWF organisations present in Europe, twelve of them in the European Union (three of the member states do not have a WWF). The process of adapting this extensive national representation into a functioning WWF European network at a policy and programmatic level is a continuing challenge.

ADAPTIVE CHANGES: THE BRUSSELS ENVIRONMENTAL LOBBY

Reviewing the list of environmental organisations mentioned earlier as having established themselves in Brussels in the late 1980s and early 1990s, it is possible to see which ones have been influenced largely, or solely, by UK groups, as with the case of WWF.

It is clear that the decision first to create BirdLife International, and then to locate a full-time office in Brussels in 1992, was largely inspired by the Royal Society for the Protection of Birds (RSPB) based in the UK. In the case of the Transport and Environment Federation, which established an office in Brussels in 1992, the role of Transport 2000 in the UK was certainly important, although by no means the only influence: the Dutch and Swedish transport and environment groups were also significant.

In most of the other cases, the influence of UK groups in establishing new structures has been less evident. Friends of the Earth in England, Wales and Northern Ireland has maintained a relatively non aligned relationship with the other Friends of the Earth groups in Europe, including the coordination office in Brussels. It has concentrated on promoting national and local interests and making use of European regulations accordingly, for instance through its effective use of Article 169 procedures to hold regional water authorities in England and Wales accountable for meeting EC water quality standards (see Chapter 14).

UK environmental groups, including FoE, have benefited from the setting up of Climate Network Europe in Brussels in 1990, but the funding and initial impetus for the establishment of the network came from the USA and from other European countries including Germany and Sweden. Likewise, Greenpeace in the UK appears to have been less influential, and less interested, in the initial setting up and operation of Greenpeace's EC unit than in the development of the separate Greenpeace International.

The conclusion from this quick overview is that, with the exception of WWF and RSPB which set up their own offices in Brussels, other UK groups have tended to operate with varying degrees of enthusiasm either through the international structures to which they belong, as in the case of FoE and Greenpeace, or through specialist networks, federations and fora such as those established for climate, transport and habitat protection. The federal structure of the EEB has continued to be used by many of the groups that also use their own specialist networks. In fact, the composition of the EEB has been largely unaffected by these developments, the main exception being the decision by Greenpeace to withdraw all its national member organisations *en bloc* from membership in 1991 out of concern over the EEB's effectiveness.

The mushrooming of the environmental lobby in Brussels prompted a review by the EEB in 1991 of its own internal working, and its relations with other national and international groups (Van Ermen 1991). The working party system of the EEB, its constitution, and its relations with international and national members were put under considerable pressure to adapt to what some saw as new competition among international groups. The needs of the environmental movement in the Central and East European countries to have some form of representation, or relationship, with groups in Brussels was also an important new factor. The internal review exercise cannot of its own be said to have directly created new working relationships among the Brussels-based groups, but it may have helped to clear some of the obstacles. There were three positive developments that occurred shortly before and after the EEB review.

First was the spur given to the four largest Brussels-based groups – FoE, Greenpeace, WWF and EEB – to co-operate more closely with each other in their dealings with the European Commission. The 'Gang of Four', as it became known, developed close working relations with the head of the NGO liaison unit in the Environment Directorate-General (DGXI) of the Commission, which resulted in regular reporting back sessions to the environmental groups from Commission officials following the normally secretive meetings of the Council of Ministers. All these groups, except Greenpeace, also received financial assistance from the NGO liaison unit in DGXI to help offset their core operating costs (Rucht 1993).

The second development was closer formal collaboration among the four groups in preparing common position statements. The first result was the document known as *Greening the Treaty* (WWF *et al*. 1991). The key demands were a commitment to sustainable development being included in the opening titles of the Treaty, recognition of European citizens' rights to a clean and healthy environment, and for strengthened integration of the environment into all the other Community policies. These demands were presented to largely sympathetic officials at the beginning of the Dutch Presidency of the Council in July 1991.

A third development was the recognition that while different groups worked on different subjects, there was an advantage from being able to pool resources with one or more of the other environmental groups for certain campaigns. WWF used this networking capacity to full effect in its 1992 campaign to reform the Structural Fund regulations which were then under review in the Regional Policy Directorate (DGXVI). The seventy or so environmental organisations which eventually signed up to the campaign representing all twelve member states were drawn directly from the member organisations of the EEB which lent its full support to the campaign spearheaded by WWF and RSPB (Long 1995).

The co-operation established by the original Gang of Four has since been extended with the addition of three new member organisations – BirdLife International, Climate Network Europe, and the Transport and Environment Federation. The 'G-7' as it is now known has produced common position statements on the Delors White Paper, *Growth, Competitiveness and Employment* (CEC 1993d; WWF *et al*. 1994), and on the 1996 Inter-Governmental Conference (Hallo 1995). The 'G-7' is also co-operating closely in its comments to the EU institutions on the proposal for a Decision on the Review of the Fifth Environmental Action Programme (CEC 1996a).

These forms of collaboration among the environmental groups in Brussels are probably the minimum necessary to gain efficiencies from limited resources. It is estimated that between them, the staff available to the environmental groups actually based in Brussels amount to no more than 20–30 persons over the past few years. Of course, there are more people working on European policy issues at the national level within each of the respective group networks. There are also other European networks with full-time staff, but not based in Brussels. Even so, the overall resources available to the environmental movement for influencing European policy making are relatively modest. They are certainly no match, in terms of numbers at least, for the other interest groups attracted to Brussels. The Commission communication on relations with special interest groups in March 1993 estimated there were approximately 3,000 special interest groups of varying types in Brussels with up to 10,000 employees working in lobbying (CEC 1993a).

THE CHANGING CLIMATE FOR
ENVIRONMENTAL LOBBYING IN BRUSSELS

Despite – or perhaps because of – some of the advances of the environmental movement in the lead-up to the Maastricht Summit (1991), the tide appears to be turning against it. Shorter term and narrowly conceived arguments of competitiveness and deregulation are in vogue; the environment and sustainability are not. The Fifth Environmental Action Programme published in 1992 at the time of the Rio Earth Summit and its very title, 'Towards Sustainability', are now being looked upon with some nostalgia by environmental groups as perhaps representing the high-water mark of the environmental advance in EU policy making. The reason is simply that the document contained commitments to targets and timetables for action that have now largely fallen out of current political debate (see Chapter 1). Instead, euphemisms such as 'shared responsibility', 'partnership', and the 'voluntary approach' are gaining ground. They threaten in many cases to become poor substitutes for an explicit environmental policy with precise environmental targets backed up with legislative force where necessary and a range of fiscal incentives where these would be more appropriate (see Chapter 3).

There are a number of worrying signs concerning the decline in the status of the environmental lobby in Brussels *vis-à-vis* the Commission. The review of the Fifth Environmental Action Programme within the European Commission between 1994 and 1996 gave only lukewarm recognition to the role of the environmental lobby. Generally, there was little consultation with environmental groups, and this contributed in large part to the unanimous, hostile reception by the 'G-7' environmental groups when the decision document on the review was published in January 1996 (CEC 1996a).

Just as remarkable was the failure by DGXI to undertake any consultation whatsoever with environmental groups before publishing a proposal for a Council decision in 1995 formalising the relations, including financial subventions, between the Commission and environmental groups (CEC, 1995d). This was in marked contrast to the attitude taken in a similar exercise by the Development Directorate (DGVIII) in its relations with voluntary organisations in the development field. In fact DGXI struck a double blow against the environmental groups. Not only did it fail to consult with them, but it also proposed in its annual call for proposals (CEC 1996b) that a maximum limit of Commission financial support for environmental organisations be established at 60,000 Ecus, a fraction of the assistance that is currently given to large federal groups such as the EEB (CEC 1996c).

Against this relatively unfavourable background for the environmental movement, there are signs that largely negative but highly targeted and effective lobbying by industry groups is gaining ground (Chapter 9). The best known example in recent years has been the co-ordinated effort to restrict the EU from introducing some form of carbon/energy tax. More recently, an alliance of automobile and oil interests is widely perceived as having been successful in weakening the 'Auto-Oil' Programme of DGXI relating to vehicle emission standards (see Chapter 3). Present signs are that even a relatively weak Commission position statement on economic incentives and disincentives for environmental protection, which one might assume would have some appeal to the industrial and commercial sector, will have difficulty emerging from the Commission in anything like a meaningful form.

Short-term competitiveness arguments are clearly in the ascendancy at this time. The extensive political lobbying structures and resources attracted to Brussels in recent years are being deployed in promoting the narrow sectoral and sectional interests of business and industry. Wider public interest issues, including environmental and social ones, that characterised policy debates in Brussels until a few years ago, are becoming more difficult to promote.

CONCLUSIONS

One of the advantages that environmental organisations enjoy is the fact that transnational co-operation of the kind that is necessary to be influential in Europe should be relatively easy to achieve, at least in principle (Rucht 1993). The reason is that the analysis of environmental problems and their causes is normally commonly shared across national frontiers. Proposals for their prevention and alleviation likewise are relatively easy to agree upon. Common goals are far less easy to identify among commercial interests where transnational and national competition is far more likely to prevail, especially within the same business sectors. This should mean that environmental groups can work together on the basis of *highest common factor* principles while some, although not all, of the interests which may be seeking to slow down environmental advance have to operate on the basis of *lowest common denominator* positions.

Set against this advantage, however, is the fact that environmental organisations appear to have difficulty in marshalling their resources and skills available at the national level and bringing them to bear on the policy formulation and decision-making processes in Brussels. This is as true for the UK as it is for other European countries. Despite the enormous national memberships of environmental organisations, popular support for the environment as an issue and an increasingly professionalised staff operating effectively at the national level, much less of that expertise and support is reflected in the day-to-day policy process in Brussels than should be the case. In other words, the environmental movement is 'punching below its weight' in Brussels despite the important advances which it has made in its institutional presence in the city over the past few years. The weakness of the environmental movement in consolidating its position and influence in Brussels during the hey-day of public concern for the environment in the late 1980s and early 1990s means that for the time being it must compete even harder. In the meantime, better resourced, and often more focused, single-issue interest groups take ever bolder positions arguing for a halt, or even a roll-back, in environmental legislation and policy at the European level.

NOTES

1 The twenty-five founding member organisations of the EEB represented all nine countries which were member states of the EEC at that time. The groups divided along the following national lines: UK (6); Belgium (4); France (4); Netherlands (3); Italy (3); Luxembourg (2); Ireland, Germany and Denmark (1).

2 Vonkeman (1994) describes the formative stages of the EEB in 1973 and 1974 as having an 'engine' in the form of Belgian national, Hubert David (the EEB's first secretary general) assisted by Geneviève Verbrugghe. Other nationalities of the founders were Belgian (Louis-Paul Suetens), Dutch (Gerrit Vonkeman), British (Julian Lessey) and other less active French and German participants. In tracing the national influences on the EEB's development, it is interesting to note that the three former secretary-generals, combining over twenty years' experience, have been two Belgians (Hubert David and Raymond Van Ermen) and one Dutchman (Ernst Klatte), tending to re-affirm the strong Belgian–Dutch links present at the outset of the organisation. The new secretary-general, John Hontelez, is also Dutch.

7

PUBLIC INFORMATION ON THE ENVIRONMENT

The role of the European Environment Agency

Brian Wynne and Claire Waterton

> The achievement of the desired balance between human activ-
> ity and development and protection of the environment
> requires effective dialogue and concerted action among part-
> ners who may have conflicting short-term priorities; such dia-
> logue must be supported by objective and reliable information.
>
> The success of this approach will rely heavily on the flow and
> quality of information both in relation to the environment and
> as between the various actors, including the general public.
> The role of the European Environment Agency (EEA) is seen as
> crucial in relation to the evaluation and dissemination of
> information, the distinction between real and perceived risks,
> and the provision of a scientific and rational basis for decisions
> and actions affecting the environment and natural resources.
>
> (CEC 1992b)

The density of environmental information in circulation has dramatically
increased in recent years. This has been true in both specialist and general public domains. In some fields, such as environmental auditing of
organisations, eco-labelling, or environmental impact assessment of
development proposals, the roles of this information are multiple, ranging
from better planning and decision making, to more environmentally
responsible decisions by consumers and investors. Information has
acquired a normative as well as a regulatory function, and the novel
importance attached to it is an essential component in the rise of a new
'civil society' (Gibbons *et al.* 1994; Melucci 1989; Beck 1992; Lash *et al.*
1996; Giddens 1990, 1991).[1]

In the past twenty years, information on environmental quality and performance has come to play a more prominent role in the implementation,

but also in the shaping of policy. The rise in EU environmental policy making is attributable to various factors. Whereas certain policies have overarching constitutional validity on environmental or health grounds, others are justified by facilitation of free trade between the member states. Often, the creation of Community-wide standards of environmental information which can set the parameters for a level playing-field in regulation is seen as a way of resolving the tension between economic integration and environmental regulation within the Single Market (Vogel 1993: 130). Environmental information is thus being put to use to serve quite different ends.

Hence the emergence of new forms of *European* environmental knowledge is both a consequence of European integration and a reflection of the different possible directions it might take. Struggles over access to environmental information and its meaning, which have become commonplace within member states, are now being played out at the European level.

This chapter focuses on some of the tensions surrounding the development of the role of the new European Environment Agency, the body formally charged with establishing a common system of environmental information for the European Union.[2] It identifies particular influences that are framing debates and commitments within the Agency as to what its core roles and relations should be. The main point of the chapter is to reflect on the implicit struggle taking shape in the emerging relations between the EEA, the European Commission and member states, over the construction of European political identity in relation to environmental policy and the new forms of 'European environmental knowledge' that are emerging.

The EEA was established by Council Regulation in 1990. Its main purpose was to set up a European environment information and observation network which would help to achieve Community aims in environmental protection. Its responsibility is to provide 'objective, reliable, and comparable information at European level, which will enable the Community and the member states to take the requisite measures to protect the environment, and to assess the results of such measures and to ensure that the public is properly informed about the state of the environment' (CEC 1990a).

Given this potentially very ambitious remit, and the ambiguities of its boundaries (for example concerning policy promotion), the Agency has been searching since its establishment for an appropriate *modus operandi*, and has had to deal with several more specific and often conflicting ideas about: precisely who its constituency should be, both in respect of users and sources of information; what the role of the environmental information it collects will be; the relationship of the Agency to the Commission, the Parliament and the member states; and how to establish the quality and effectiveness of information to be used for policy. In an increasingly information-dense arena, the EEA has had to define its relations and boundaries quite stringently, to avoid duplication, waste and confusion. But to define 'environmental information of European significance', for example, in order to clarify relations with local and member state actors, is a complex matter with important ramifications for information and public policy.

Unquestioned and largely conservative pressures on the relationship between European institutions and the member states are playing powerfully formative roles in defining the way that the Agency will operate and who its key actors will be. Similar assumptions and tensions were at play also in the EEA's predecessor – the Commission's Co-ordination of Information on the Environment, CORINE.[3]

Various other assumptions and 'framing' commitments also have had significant influence on the EEA, and on information and policy more generally. One of the most prominent of these is harmonisation at the European level, which often means standardisation, enshrined in the obligation of the Agency to provide 'objective, reliable and comparable' data. In environmental regulation generally, a key tension is that between the setting of European standards and the application, particularly since the Maastricht Treaty came into force in 1993, of the subsidiarity principle. On the one hand, European standard setting, coupled with the dissemination of monitoring results, has made previously closed policy communities, such as that to do with water quality, more accessible and responsive to public and pressure group influences (see Chapter 14; Ward, Buller and Lowe 1996). On the other hand, in the information sphere, standard setting at the European level can have the opposite effect and risks alienating wider civil involvement if carried out in a manner insensitive to local and cultural concerns. Standard setting means the setting of norms, and these are not only technical or scientific in nature, but involve ethical judgements too. It is an open question whether the Agency is prepared for handling such issues.

Whereas other chapters here address the issue of Britain's adaptation (or reaction) to EU policy culture, we look more closely at that environmental policy culture. One of our main observations is that the wider 'information culture' in which the EEA exists poses completely new challenges for the Agency and the formal policy institutions; and it poses questions which should encourage re-examination of traditional member state relationships to EU bodies. As a much more pluralistic policy framework comes into being in European environmental policy, the provision and role of information for policy embrace a much wider set of actors, posing concerns about what information is for, and which institutions, groups and individuals should provide and use it. These concerns, in turn, tap into a wider debate about EU institutions and the process of Europeanisation at a time when the precise development of the Union is under considerable negotiation. Governments are the key actors in this drama, although in their struggle to define a manageable order, their centralising reflexes coincide far more with than they diverge from those of EU institutions. The tensions we describe seem part of a much larger historical development in which traditional modes of authority and political legitimacy established around 'objective knowledge' are being re-negotiated.

We turn now to give a short background to the European Environment Agency, before considering the contemporary role of environmental information in the public and policy spheres.

BUILDING THE EUROPEAN ENVIRONMENT AGENCY (EEA): BACKGROUND

In the late 1980s, when European public support for environmental policy was perhaps at its peak of political expression (Vogel 1993), green MEPs were keen to establish a European equivalent of the US Environmental Protection Agency, with powers to initiate regulation, as well as full powers of inspection, monitoring and enforcement. This would have marked a radical step towards surrendering national sovereignty in such matters, and even some green supporters thought it risked encouraging a nationalistic backlash against the idea of 'foreign' inspectors thought to be meddling in sensitive national issues. Nevertheless, the wish to capture the strongly optimistic environmental mood in some European institutional form was virtually irresistible and, with strong parliamentary backing, Commission President Jacques Delors introduced a Regulation to set up the European Environment Agency as an organisation to provide, for European bodies and the public at large, 'objective, reliable and comparable information' on environmental quality. To avoid opposition from member states, it would have no formal regulatory role, not even in inspection and enforcement, and it relied upon the idea of objective (though policy-relevant) information to define a neutral role in relation to the Commission's prerogative to initiate policy.

The new 'stand-alone' agencies of the European Union, such as the EEA,[4] have no constitutional precedent to help delimit significant ambiguities in their political position. They are not Commission bodies, but have an independent constitution. The EEA has an independent Management Board comprising a representative from each member state (including two non-EU states, Iceland and Norway), two from the Commission, and two experts designated by the European Parliament. The importance accorded to the EEA is reflected in the fact that each member state's representative is the senior environmental policy official in that government, and that the Commission's members are the two most senior officials in DGXI (Environment) and DGXII (Science, Research and Development). The Board also elects its own chair, giving it an added dimension of independence.

Although the Council Regulation (1210/90) (CEC 1990a) setting up the EEA was formally signed in May 1990, failure to agree a location meant that it was not actually brought into operation until December 1993, by which time a Danish site (in Copenhagen) had finally been settled on. The immediate priority, besides appointing staff, was to agree and get under way a work programme consistent with the general principles set down for the Agency in the Regulation. The EEA's structure is important in this connection. The structure envisaged for it was decentralised, with a distributed Information and Observation Network (referred to as EIONET) which was to consist of three parts: existing national information frameworks; the national 'focal points'; and EEA 'topic centres'. The last are scientific institutions contracted by the EEA to execute particular policy-oriented information-gathering tasks on relevant topics such as air quality, water quality, soils and nature conservation. 'Focal points' are the national environmental authorities (normally the Environment Ministry of each country – in the UK it is the Department of the Environment).

This 'decentralised' structure was established by member state representatives. To a significant extent, the components of the network represent the official national environmental communities through the selection of what governments perceive to be the most important bodies and institutions. This has left the role of voluntary organisations, local government agencies, and wider civil (and even scientific) involvement quite markedly marginalised. Although the Agency describes itself as a 'distributed' network, in fact its most active parts consist of a relatively restricted network, composed largely of 'official' bodies and spokespersons, in contrast to the variety of actors and institutions involved in environmental protection across Europe. The selection of specialist topic centres responsible for cross-national data collection proceeded on the assumption that each member state should have one, and considerations of political diplomacy (which country should host which topic centre) were as important as those about qualifications and expertise. The reinforcement of certain official channels is also illustrated by the national focal points: their very responsibility has meant that they act (often inadvertently) as gatekeepers to national activities on European information, throughout the EIONET structure. In this way, they also shape the future identity of the EEA, determining whether it develops as a centralised, or perhaps a more broadly based, European body.

It is a central principle that the EEA should provide 'objective information necessary for framing and implementing sound and effective environmental policies' (Regulation 1210/90, Article 2(ii)) (CEC 1990a) – that is, information of use to public policy – and this ethos has been pursued by its Director, explicitly distinguishing this role from scientific research. In a publicity pamphlet issued in 1994 he stated:

At the EEA, we have no use for data amassed for the sake of completing an elegant yet arbitrary pattern, or in case they might some day come in handy to somebody. Our job is to put data to work, to deliver information products and services based on more efficient recovery of the existing data backlog and on determined expansion of future options for applying information and its technology to tackle real problems.

(1994: 1)

But formally the EEA has no policy role, beyond simply furnishing information to policy-making and other bodies. In practice, however, the boundary between 'neutral' information and policy shaping is ambiguous, and changing. This has been reflected in the relationship between the EEA and the EU body actually charged with environmental policy – the European Commission and its DGXI – where the potential overlaps in responsibility require careful negotiation. As experience in the EEA has already shown, there are various ways in which 'objective, reliable and comparable information' can exercise policy values. As regulatory cultures change towards more information-dependent styles, the policy importance of an information agency is likely to grow. This potential is amplified by the fact that the EEA is charged in its Regulation to provide not only environmental-quality information but also information on the upstream pressures and stresses, such as the polluting activities, that affect environmental quality. The implications for existing institutions, especially the formally constituted policy bodies of the EU and member state governments, are considerable, not least because they mark out the contours of the contemporary struggle to create a politically legitimate and democratically rooted environmental policy culture in Europe.

ENVIRONMENTAL INFORMATION AND PUBLIC POLICY

The EEA's remit establishes an overarching logic which suggests that information will enable the Community and the member states to 'take the requisite measures to protect the environment' (Regulation 1210/90, Article 1.2). In the 1994 EEA publication *Putting Information to Work* the normative shaping power of information is again stressed: 'Policy makers may direct information agendas, but the process can easily be reversed if, for example, data emerging from particular investigations or assessments reveal a need for a complete change of policy direction' (EEA 1994: 4). Information is thus seen as having an active normative role in its potential to open up possibilities for policy that previously did not exist.

Over the past twenty years public information has become much more prominent as a strategic element in policy making, as public authority 'command and control' regulation in some areas of environmental policy has lost favour, and has been replaced by market mechanisms such as pollution-permit trading and voluntary agreements between governments and industries. Newer 'more flexible' and decentralised approaches to environmental protection and regulation need information on environmental performance, of companies or products, to be more widely available, as a direct instrument of policy evaluation. This trend is consistent with a kind of 'private sector' philosophy within the policy sphere, particularly visible in the monitoring of policy implementation (Ludlow 1991: 107–8).

Public information on environmental performance and quality is becoming like a policy instrument, in that it will supposedly help create political pressure to introduce focused policies, or implementation regimes, to improve matters highlighted by that information. Standardisation is implicit in this process and may arouse the sensitivities of member states (Rhodes 1986: 29; Marks 1992: 217). The tensions that can result were well illustrated when the draft of the 1995 Report on the State of Europe's Environment, commissioned by the pan-European ministerial meeting at Dobris Castle in 1989 (Stanners and Bordeau 1995), was presented to the Management Board of the EEA in February 1994. The comparative information on environmental quality in different European countries, which the Report provided for the first time, provoked defensive demands from several national representatives to be allowed to vet the draft for politically embarrassing information. Editorial control by member states was refused, with reference to the ministerial auspices of the Report, but the episode starkly exposed the sensitivities and pressures.

More generally in European policy development, Rhodes has noted a 'recurrent problem' in 'the tension between the need for information and expertise, especially on the implementation of policy, and the determination of national governments to reserve supranational negotiations unto themselves' (1986: 29). DGXI is at the heart of this tension, being dependent on the very actors it is monitoring (namely member state governments) for the information it needs to evaluate national performance. Although informal channels of information have developed between local and national groups and DGXI (Krämer 1991), this has been very patchy, and sometimes has been strongly resisted by member states. Such channels have been particularly well used by UK environmental groups and one of the reasons for the UK Government's enthusiasm for the EEA is to shift the spotlight which it feels has been unfairly focused on its environmental performance (see Chapter 3).

While public information has been taking on a more strategic role in environmental policy, the nature of the policy process itself, and not just the instruments used, has also been undergoing profound changes. As many other chapters in this volume point out, a shift has taken place (in the UK, at least), catalysed in part by European requirements. Policy and regulatory systems in place today rely less on in-house expertise and closed consultation, negotiation and accommodation of specific concerns, and are more oriented towards transparency of performance against explicit and formalised standards. The formalisation of the role of environmental information, through monitoring and reporting procedures, has indirectly led to greater public participation which has brought that information forcibly to bear upon the policy processes and opened out regulation to a much wider and more fragmented community of interested parties.

At least in countries such as the UK the policy process is gradually being superseded by a much more open, pluralistic and conflictual model in which the authority of official policy bodies and scientific expertise is regularly challenged (but see La Spina and Sciortino (1993) for comments on important differences between Mediterranean and other member states). Although public trust in official policy bodies and other established experts has receded markedly, such bodies are now required to supply more information, partly in an attempt to stem rising public mistrust and disaffection. At the same time, other sources of expertise – such as independent scientists and non-governmental organisations – have apparently become surrogates for public support and identification (Princen and Finger 1994: 223). Scientific sources of information and authority have also multiplied, and the media have assumed an important role in encouraging a greater place for non-official arguments and perspectives. Environmental groups have developed their own specialist expertise and have been drawn into national and international policy-making fora as the public legitimation of formal policy bodies has come more into question (see Chapters 5 and 12).

In this development of an unofficial 'civil society' with its distinct but highly pluralistic and fluid patterns of solidarity, identification and value, information has become more dense, more central, and more broadly cultural, but at the same time more contested and problematic (Gibbons et al. 1994). The relations between formal policy bodies and their somewhat overstretched constitutional mechanisms on the one hand, and this informal, ill-structured and fluid but culturally rooted 'civil society' on the other, are complex but not necessarily in competition. Much EEA literature and rhetoric signals at least a subconscious acknowledgement (and often endorsement) of 'extra-official' involvement in information provision and environmental policy making. Some, such as the representatives of the European Parliament, have argued that this more openly demanding and broad-based role and identity (for information) has an instrumental justification at least, as the only means of rescuing the political legitimacy of formal policy institutions. As we will go on to illustrate, these pressures to reflect contemporary trends in the pluralisation of public policy reveal the tensions surrounding the Agency.

FORMATIVE INFLUENCES IN THE EEA'S ESTABLISHMENT

In the past three years, the EEA has had to invent itself, guided by its founding Regulation. The changing political context has broadened the scope for interpreting and responding to its remit. A crucial factor has been the Management Board's interpretation of the relationship between the EEA and the Commission (and particularly DGXI), and the official bodies of the member states. The effective authority and policy scope of the Commission have always been more fragile and dependent than its formal constitutional position might suggest, but have become more so in recent years, compounded by projections of EU policy as infringing on legitimately sovereign concerns. The EEA can be seen to have overlapping responsibilities with the Commission. Though a possible source of criticism, this could equally provide opportunities for synergy that would strengthen environmental policy. In particular, the EEA's relative independence from formal Commission processes should, in principle, allow it to operate in a way that would complement DGXI's more circumscribed abilities. Yet it appears that such opportunities may not be taken up.

A key issue is who are the intended audiences for the information that the EEA is to provide. This is a matter of some ambiguity and conflicting interpretation. There is no question that the Commission is a special partner (Regulation Article 2 (ii)), though here there are disagreements about the extent to which it should act as gatekeeper in this relationship. The DGXI representative on the EEA Management Board has asserted a strict interpretation of Article 1.2 of the Regulation to insist that the EEA's proper user relationships are limited to the Commission and the member state governments only.

However, there are several counter-arguments to this interpretation. For a start, references in the Regulation to 'the Community' are ambiguous. Even if they mean the institutions of the Community rather than the Community as a whole, this should include the Parliament, the Committee of the Regions and other bodies as users, not just the Commission. Second, there are other Articles (e.g. 2 (vi)) which explicitly require the EEA 'to ensure the broad dissemination of reliable environmental information' and to make data emanating from the agency accessible to the public (Article 6). Third, other specified tasks of the EEA – for example, ensuring reliability via quality control of environmental information (Article 2, (iv)), or stimulating exchange of information on clean technologies (Article 2, (ix)) – can only be fulfilled by direct involvement with diverse, specialist non-governmental users and information holders, such as environmental groups, trade associations and local government bodies. Furthermore, this more direct and engaged vision of the EEA's role in civil society is encouraged by the recognition in such EU policy commitments as the Fifth Environmental Action Programme that environmental policies for sustainability can only be delivered by engaging the diverse participation of the widest social partnerships.

Even though the EEA's own publicity specifically refers to the role of experts from 'research and education fields and partner organisations, professional organisations, NGOs and industry' (1994: 7) there remains a tendency to narrow down definitions of responsibility and partnership within the Agency's Management Board. Often in its discussions, 'Community' is taken to mean the Commission. This indicates that the EEA is subject to many of the familiar, more informal codes of conduct and expectation that have become enshrined in Commission politics over the years (Andersen and Burns 1992). This was a central concern of the European Parliament back in 1990, when consulted on the EEA Regulation.[5] Andersen and Burns have noted that although policy processes may be quite open in European institutions, outcomes are not equally open: 'Institutional features frame and limit the search for possible outcomes.' They point to strong pressures stemming from the 'norms of rationality in the decision points'. This appears to be what is happening in the process of defining the EEA's legitimate partners or 'users'. The traditional instinct within the Management Board, and of Commission officials, is to close down the potential reach of such partnerships, centring them markedly on the relationship between the Agency and DGXI. Such tendencies are perhaps not deliberate control strategies so much as entrenched assumptions about the public policy process that do not acknowledge the changed situation. If the EEA were to acquiesce, it could incur the same problems of public alienation and lack of identification which cripple existing formal policy institutions from advancing policy commitments that engender positive public support.

THE EEA AND THE MEMBER STATES

It is the member states' representatives who are the key actors in this struggle to define the identity and ethos of the EEA, and indirectly the political culture of Europe itself. The designation of environment ministries as the 'national focal points' ensures political control as well as necessary co-ordination, but potentially inhibits free access and exchange between the EEA and independent bodies at sub-national or international levels which may be in conflict with member state governments. Such control of data flows was a concern discussed in a recent inquiry by a House of Lords Select Committee on the EEA. When the Chairman of the Committee suggested that there was likely to be some 'filtering' (1995: q. 24, p.14) of information by the Department of the Environment, Mr Osborn of the DoE answered, 'I should like to put all your minds at rest if you think that we are going to act as some sort of censor on information going to Copenhagen.' At the same time it was acknowledged that issues of data quantity, data quality and 'unmanageable data flows' would have to be co-ordinated in the UK (through the DoE), as in other countries. On the issue of accessibility to EEA networks it was asked whether 'unsolicited data would be verified by the national focal point from where it had come' (q. 35, p. 16). Mr Martin (of the DoE) answered:

> the Agency's role will be to bring together data and to advise the Commission and to do that it needs to have reliable information and if they are unaware of its origins or how it was collected and that sort of thing then I think that it is necessary for them to come back to the national focal point to try to verify how good that information is because we do not want them to go to the Commission with bad information and the Commission basing their policy advice on bad information.
>
> *(ibid.)*

The DoE acknowledged that at times it would therefore have to verify the 'authenticity' of data from unsolicited sources (q. 38, p. 17).

When asked how the Director of the EEA would deal with information that was politicised by vested interests ('either a national vested interest to show perhaps their bathing water is cleaner than some of us might have experienced it; or the non-governmental organisations, the specialist lobby groups; or perhaps the French, who would like us to think that there might be more turtle doves in the Dordogne than perhaps our own [Royal] Society for the Protection of Birds might believe to be the case' – q. 6, p. 2), Señor Beltran replied, 'We have been told that the Agency is a watchdog with no teeth and I always say that one of the best teeth now is to have written information. If you produce good information the public and the administration will do the rest.' Such (tautological) statements perpetuate confusion as to how to ensure the quality of data, which is a genuine problem for the EEA. In practice, within the workings of the Agency, the most effective and rational means of trying to ensure quality of data, and in deciding what role to play in information conflicts between different bodies and authorities, is through the controls exercised by the national focal points, whose position is, in any case, sovereign.

One official of a national focal point reacted to the idea of relaxing this gatekeeper role as follows: 'We don't want any eccentric scientist just getting in touch with the Agency themselves and establishing their own relationship.' Such reactions are consistent with other research looking at the role of sub-national actors within the EU. Marks (1992: 217) has noted the 'threat' posed of a loss of formal control: 'Once policy networks linking subnational governments to the EC have been created, there is no certainty that they can be dominated by national government.' The information that may be excluded as governments seek to retain their control is not all necessarily dissentient: as Ward, Talbot and Lowe have found (1995), useful and relevant expertise in statutory agencies, for example, has failed to permeate to the European Commission, due to the DoE's overarching role in mediating official links with the EU. We have noted similar blockages in the EEA's predecessor, the CORINE programme, where information handed into the Commission through official networks was a very poor reflection of the available scientific data in member states at the time, largely because of restricted access to data networks (Waterton and Wynne 1996; Waterton *et al.* 1995).

Undoubtedly, there are pressures on member states to avoid sheer incoherence and overload, as well as to control communications and relationships between the EEA and other bodies within that state's sovereignty. In the House of Lords Select Committee report such issues are referred to as the 'control of the quality, quantity and access of data flows' (7). Yet it is unclear who should be making the assumptions and judgements about quality, quantity and – perhaps most crucially – access. The issues of trust (which institutions can be trusted to provide authentic data?) and of what constitutes *bona fide* data have yet to be introduced into discussions about data networks, gatekeepers and flows. So, although the Agency is committed to 'rationalising' already prolific data sources, it remains unclear as to how this will be achieved.

A familiar response to expressed fears of centralisation is that new information technology will enhance circulation and access, and counter any centralising tendencies. However, such systems may actually inhibit the opportunity for potential participants to identify and evaluate the trustworthiness of information sources. By restricting conflict and debate, the Agency may, paradoxically, be showing signals of a dangerous and ultimately weakening insularity. It may thereby reduce the risks of providing ineffectual, inappropriate or even inaccurate information in processes of co-ordination and standardisation (Porter 1995; Jasanoff 1996). But by maintaining control through singular national channels within the EEA network, it is potentially at odds with the norms of an open scientific network which honour independent channels of cross-reference, criticism and debate.

RATIONALISATION AND STANDARDISATION OF INFORMATION FLOWS

Rationalisation is understandably a central interpretation of the Agency's task as a 'screening' and 'quality control institution' (evidence of Señor Beltran to the House of Lords Select Committee 1995: 2) At the same time rationalisation of data introduces the problem of maintaining respect and sensitivity to legitimate variation in data flows and sources. The impetus to rationalise and standardise data is amplified by the severe pressures on officials bombarded with overlapping demands for environmental information from an increasing number of international bodies, including OECD,

Eurostat and the various Directorates of the European Commission. To these can now be added the several separate topic centres of the EEA. The potential duplication and waste are enormous and growing, at the same time as the available administrative resources are shrinking.

The need for sensitivity to local difference still remains, however. For example, some federal governments such as Germany and Spain have asserted that demands for data need to be sent to each *Land* or province to reach the constitutional bodies responsible. This has been viewed on the EEA's Management Board as creating a further layer of perceived 'inefficiency', and has added to pressure for centralisation and co-ordination in order to meet 'efficient' standards of information processing. Yet such rationalisation of effort may squeeze out distinctions of purpose and meaning (thus quality and usefulness) which different information requests may embody. The pervasiveness of the dilemma of how to rationalise data without losing local validity and relevance is not yet recognised.

Alongside rationalisation, the Agency's responsibility to standardise European environmental information, to develop data and information (in the words of the EEA's Executive Director) 'to the level of uniformity required for analysing the big picture and setting benchmarks for implementing agreed measures or legislation over wide areas' is one of its most important roles (Jimenez-Beltran 1994: 1). 'Calling to order' existing information resources, defining what is information of 'European significance' and what is not, deciding which information is appropriate for use by planners, policy makers, and so on, aggregating and representing information to facilitate (if not enforce) change – all these activities involve creating standards which will reinforce certain perspectives and marginalise others.

For example, in a study examining the way that the Commission created a standard classification for natural habitats defined to be 'of Community interest', we found that the resulting 'standardised classification' (called CORINE Biotopes[6]) has limited credibility outside Commission institutions for its reliability. Although it played a vital role in Community policy enabling the Habitats Directive (92/43/EEC) to be drafted, many national and sub-national actors do not recognise the classification (or the relevant annexes of the Directive) as providing an accurate account of the nature conservation resource needing EU-wide protection. This may well jeopardise the implementation of the Directive.

Looking behind the process of creating this classification, we have examined some of the reasons for its lack of acceptance by the scientific and nature conservation communities beyond the Commission. The influences that we identified are pertinent also to the EEA – especially to the way in which it collects, 'screens' and harmonises (or creates standards in) information. Information quality in CORINE Biotopes was hampered by the following factors: a restricted network of actors involved (largely influenced by Commission and member state interpretations of 'subsidiarity' in data collection); a presumption that information had to be collected in line with policy time-scales (that is, too fast for the information to be properly validated); and a very restricted conception of the users of such a classification (that the implicit 'user' was the Commission, not the wider scientific community or public in Europe). These influences combined to alienate, rather than to enfranchise, the very community that is now enacting the legal requirements of the Habitats Directive.[7]

The example of CORINE Biotopes is typical in that it illustrates a prevalent tendency within the Commission to see standardisation issues as largely technical in nature.[8] Yet, the assumptions built into the way that data were standardised were largely political and policy oriented, and the resulting classification clearly reflects their influence: in that respect CORINE Biotopes is a classic example of a cultural product of the Commission. A further example concerning the harmonisation of standards for the transboundary movement of toxic waste management and regulation in the Community illustrates a similar point.

Before the Single Market, each member state had its own approach to waste regulation. In particular the systems of defining wastes (which required different forms of special handling, treatment and disposal, with different cost implications) varied from one member state to another. This meant that a consignment could move from one country where it was regarded as a special waste needing careful registration, monitoring and handling, through another country where it was not deemed a special waste and where it was not controlled. Several such cases, where special wastes designated to a final destination were lost in transit and were assumed to have been dumped, led the authorities to recognise the problems involved in this lack of uniformity. It was assumed that the Single European Market would include a single market for waste movements, treatment and disposal, gaining efficiencies of scale and honouring commitments to free trade, but this required a single set of waste classification criteria which would have to operate within a system of cradle-to-grave control all over Europe (see Chapter 11).

In practice, moves to harmonise European waste classification systems proved far more difficult than the Commission envisaged, leaving the growing commitment to waste trading across Europe uncontrolled. Analysis showed that different 'local' technical criteria in member states for waste classification reflected tacit differences of institutional arrangements, political cultures and purposes (Wynne 1987; Laurence and Wynne 1989). For example, the Dutch insisted on precise, chemical-constituent criteria which were rigidly enacted, whereas the UK used very imprecise health properties criteria. But these differences reflected the distinct ways in which local and central authorities interacted with waste producers and scientific advisers, as well as different distributions and levels of trust among the different actors involved. Discretionary judgement and flexibility were markedly different

132

between the two countries as part of their distinct political cultures and institutional arrangements. Harmonisation of the technical criteria would have almost entailed harmonisation of these institutional and cultural differences, which, of course, is a completely different order of problem. Although initially the officials responsible for waste policy did not recognise this deeper structural dimension of a supposedly 'technical' standardisation issue hampering the free market, ultimately a policy change with a presumption against the transfrontier shipment of waste was brought about.

Both the above examples serve to illustrate a 'European' tendency to emphasise the technical, universal and standard, rather than the cultural, variable and local. Unrealistic expectations about the necessary supporting conditions for common technical criteria can result in a particularly superficial form of standardisation which creates a gulf between 'standardised' definitions at the European level and quite different understandings of what is realistic and practical at more local levels. A variety of studies in the sociology of 'nature', science and technology have shown how the universality of scientific concepts and methods has required enormous 'invisible' work in re-negotiating social practices, training, even the identities of participating actors, in order to render the standardised form valid and effective (see, for example, Latour 1995; Porter 1995). Since rationalisation and standardisation are among the most prominent tasks of the EEA, these issues of local variability and the question of how to respect legitimate cultural differences while maintaining enough consistency across cultures to provide a comprehensive overview are among the most difficult challenges facing this new institution. As standardisation is cultural and has normative force, it frequently involves political and ethical judgements. These important dimensions, however, tend to remain buried under a technical rhetoric of standards and rationality.

CONCLUSIONS

This discussion of the EEA's place in EU environmental policy and the relationships between member states and EU institutions has been built around two basic issues: what tacit models of public policy are being exercised (and reinforced) in the formative commitments shaping the EEA? And what is involved in information harmonisation?

We noted earlier that standardisation in the environmental policy field can be viewed in opposite terms. In Chapter 14 Ward describes the constructive policy consequences brought about by standardisation at European level of normative technical criteria for environmental quality. He argues that improvements in drinking water and bathing water quality would not have been brought about in the absence of the European standardisation of these norms. It is therefore worth reflecting upon the forms of standardisation which we are analysing in the environmental information sphere, to ask how they might differ from those forms which Ward discusses.

The standards which Ward analyses – quality or performance standards – are expressly normative, and they are sufficiently transparent and precise that actual environmental quality can be measured and compared with them, to allow clear evaluation of the policy performance of any member state. Leaving aside the origins and rationale of the normative standards, once they come into being through the institutional mechanisms involved, they have political force, and can be used to exert effective pressure.

The forms of standardisation we are analysing in the information domain are different in significant respects. They operate at a more 'upstream' technical level, where the choices and commitments are much less visible. They are buried in the origins of what is presented (objective environmental information) as merely discovered or measured, as if objectively existing in nature without human intervention, rather than as in the other case, normatively negotiated and chosen. Information standardisation often takes place in the classifications which define comparable entities before any measurement even takes place; thus it is far less visible and accessible to evaluation than overtly normative quality standards. It is not possible for the implications which standardisation may carry for the neglect or subordination of variations and local differences, which may be significant to particular social and cultural groups, to be recognised and debated. In the CORINE Biotopes programme, for example, habitat differentiations that are important in some sub-national areas were submerged into single elements of the standardised 'natural' classification.

These effects of standardisation in information classification and bodies of data are less visible because they are more 'black-boxed' in the very framing of knowledge that is thought simply passively to reflect natural reality, but which actually incorporates normative value commitments or implications. This less accountable and transparent kind of standardisation had its equivalent in the case of the water regulation analysed by Ward (1997), in that the Bathing Water Directive, in addition to setting quality standards, required that the 'bathing beaches' to which these standards should apply be defined. The partiality of the definition employed by the UK Government was at first not at all visible in the public policy domain.

The point here is not whether standardisation of such basic classifications is to be welcomed or opposed. Clearly in many circumstances standardisation introduces extra power, scope and clarity into policy methods and data, which outweigh the possible negative effects from the inevitable dissolutions of variation (Porter 1995). However there is a need – unfulfilled so far – to recognise what is involved when this is conducted for European environmental information, and to encourage wider awareness and acceptance of the processes and their implications.

Information harmonisation at the European level – in which the EEA is absolutely central – is important and constructive, so long as this challenge is conceived as one in which the basic categories of information are a combination of objective factors and human commitments which lend them constructive meaning. Ladeur (1996) rightly questions the conventional sharp distinction between information (seen as objective 'fact finding') and decision making, where normative commitment is recognised. In reality, knowledge production and decision framing occur together, in a process of mutual construction that combines normative and informational inputs.

Furthermore, and consistent with this development, public authority, legitimacy and thus effectiveness of knowledge and policy can no longer be seen as deriving only from formal models of constitutional representative politics and administration. Sources of credibility and legitimacy are much more diverse and independent of formal institutions, arising in variegated informal networks and movements which are indeed often inspired by express alienation from 'official' processes. This has been especially evident in the environmental sphere. However, we do not treat these public networks of identification and moral energy as alternatives to traditional formal institutional policy structures, but as a cultural fabric of a new, more complex and decentred politics with which such formal institutions need to come to terms if they are to rescue existing policy making from widespread and profoundly serious public disaffection.

The process of harmonisation should thus be seen as a collective learning process, not directed towards an objective pre-existing truth, but a mutual learning between the local and the universal, and between natural and human categories. Jasanoff has captured this principle in her discussion of harmonisation as reasoning together:

> Standardisation can be recognised as a fruitful site for exchange among competing views of knowledge, politics and action – not merely as a bureaucratic procedure to facilitate convergence towards some determinate, technically calculated endpoint. The power of harmonisation would flow importantly in this model from its capacity to reframe problems for collective solution, to extend the boundaries and parameters of relevant knowledge, and to translate experiences gained in one socio-political context to the needs and circumstances of others. Harmonisation would become more explicitly a vehicle for parties with divergent views about risk to reason together.
>
> (Jasanoff 1996)

Information harmonisation at the European level is a crucial and in principle constructive feature of the EEA's role. Yet, as we have seen, this involves much more than mere technical adjustment of locally or nationally varying systems of designation, definitions or specific criteria. The examples of the CORINE and toxic waste definitions showed the institutional and cultural ramifications which are typically neglected in the dominant assumption that such technical criteria are entirely divorced from their context. They also raised the (often buried) question of what standardisation is designed to achieve. On the one hand, standardisation of nature conservation practices across the Community for sites 'of European interest' should not have obscured the question (at least for bodies like the EEA) as to whether member states in the EU should retain sovereignty over nature conservation in their territory. On the other hand, standardisation of definitions of waste to facilitate the functioning of the Single Market should not have prejudged the question as to whether free trade in wastes would have environmentally desirable or undesirable knock-on effects.

We see a genuine dilemma here for the EEA. On the one hand, rationalisation and standardisation are central and necessary tasks; on the other, the EEA is under immense pressure – part born of narrow, outdated imaginations of its place in public policy, part born of sheer practical mismatch between limitless expectations and very tight resources – to compile and harmonise information in a way which divorces it from its social and cultural contexts, thus rendering it ineffectual and lacking legitimacy to anybody except the Commission, but by the same token implicitly insulating the Commission from wider political and cultural identification. The tendencies apparent in emergent EEA 'networks' for member state officials to act as central gatekeepers of information flows between the wider reaches of Europe in those countries and the EEA, underline this problematic trajectory.

The main point of this volume is to address what has been the effect on UK environmental policy and institutions of its membership of the EU. In this chapter we have thrown a different perspective on this general issue by focusing more broadly on the EEA's emergence in the Europe-wide changing policy culture, in which information is coming to play a more strategic role at the same time as it has become more diversely sourced, more pluralistic and more intensely contested, along with its legitimacy being wrested from the sole privilege of formal policy institutions. We have noted that this cultural politics into which the EEA has been pitched contains crucial dimensions of public alienation from conventional policy, yet it exists alongside forces for centralisation and formalisation that reflect outdated assumptions and related pressures. Thus the EEA's potential for regenerating some public legitimacy and effectiveness for the Commission and member state governments is great, but is being undermined by lack of awareness of these as issues and dilemmas which the EEA has to face. At national and sub-national levels the EEA could avoid a marginalised and ineffective role if it were allowed by its member state controllers to make its own effective partnerships with relevant providers, critics, validators, users and sources outside official government and EC channels – not at the expense of the latter, but on the contrary, to their ultimate benefit.

NOTES

1 See Wynne (1996) for a discussion of the latter three authors' interpretation of this phenomenon.

2 The chapter draws on Brian Wynne's experience as one of two European Parliament nominees on the EEA Management Board as well as on policy-oriented research (carried out for an environmental pressure group (WWF-UK), and for the Economic and Social Research Council of the UK).

3 CORINE (Co-ordination of Information on the Environment), was the Commission's first environmental information programme. It ran from 1985 to 1991 within DGXI. (CEC, 1991a and 1991b)

4 Other new European agencies include the European Training Foundation, the Office of Veterinary and Plant Health Inspection and Control, the European Monitoring Centre for Drugs and Drug Addiction, the European Agency for the Evaluation of Medicinal Products, the Agency for Health and Safety at Work, the European Monetary Institute and the Office for Harmonisation in the Internal Markets (Trade Marks and Designs).

5 The Parliament commented: 'As the proposal stands, the Management Board is a sort of mini Council of Ministers with all the technical, organisational and institutional problems that this implies. It would be quite unable to do justice to the Agency, given the wide-ranging and urgent tasks with which the latter is entrusted' (European Parliament Session Document, 6.2.1990, Doc/EN/RR/82228, 6).

6 Biotopes is here an alternative word for the more commonly recognised term 'habitat'.

7 In practice, the conservation policy community are relatively pragmatic and are concerned with getting on with the Directive (they see it as an important Directive for conservation), but it remains to be seen quite how the interpretation of the Directive's annexes (based on CORINE Biotopes) will work out. In some cases, member states might exploit the fact that the annexes make minimal demands on them. In other cases, pressure groups and conservationists will try to expand the definitions of the annexes, conserving more than is strictly required. Both responses can lead to problems in the legal context, as the way that the Directive is implemented will be compared to the precise wording of the Directive, including the technical definitions of the annexes. The lack of consensus on the validity of CORINE Biotopes as a basis for Annex 1 is quite likely to lead to controversy.

8 This was evident in the whole of the CORINE programme. Problems in harmonising data were attributed mostly to technical updating problems, when they often were, to a large extent, the result of much more complex social, political and cultural tensions and difficulties. See CEC, 1991a and b. Or for a discussion of this in the 'Biotopes' context, Waterton et al. 1995; Waterton and Wynne 1996.

8

LOCAL AUTHORITIES

Janice Morphet

The relationship between local authorities and the delivery of environmental policy is a long established one. Indeed, the local government reforms of the 1880s had at their heart the need to promote public health standards and public safety. The relationship with Europe also stretches back over many years – initially, in the post-war period, through twinning arrangements and, subsequently, through interest-based networks which have developed, particularly over the past ten years.

Local authorities have become increasingly aware of the role of the European Union in their work, primarily as a source of funds to support initiatives to deal with local economic restructuring. The squeeze on local authority finance over much of the same period as Britain's membership of the Union and the availability of this alternative funding source have encouraged the notion that the EU's main relationship with local government has been in this funding role. A new breed of staff has emerged whose job it is to act as intermediaries in this process, including some who are based in offices in Brussels.

However, as this chapter will show, local authorities have also become increasingly aware of the central role of European environment policy in their own work, particularly since 1990. The chapter puts this growing realisation into context, looking at the expanding links that local government has with European institutions, and the involvement of local authorities in European networks. It then examines the implications of these developments for the relative power and autonomy of local government and the scope for local authorities to seek influence in the policy process in addition to resources. The chapter concludes with an indication of the emerging opportunities for local government in European environmental policy.

THE CHANGING RESPONSE OF LOCAL AUTHORITIES

Local authorities are experienced in understanding and shaping issues of critical importance to their localities. Over the past hundred years it is in dealing with environmental matters, including the application of regulations and the provision of services, that local authorities have been able to respond to the express differences between them. For many years, central government regarded the delivery of such services as a matter of local concern and left local authorities to pursue their own goals.

However, over the past twenty years, there have been increasing pressures for central government to intervene. The first of these has been the growing role of the European Union in the harmonisation of environmental standards and regulation. The second has been the increased requirement for competition in the delivery of services (a requirement which, although part of the world trade regime, has actually been delivered through the means of EU law). When the UK first joined the EU, central government introduced European legislation through the application of domestic legislation but, as a later section will demonstrate, even that smokescreen is no longer available to either Parliament or the UK courts. There has certainly been an increasing understanding on the part of local government officers of the role that Europe plays in framing the rules which they operate. Much of this developing understanding has arisen from the experience of local authorities in attempting to lobby Parliament or the DoE for changes in proposed environmental legislation, only to discover that the stage at which domestic consultations take place is already well beyond the point at which any changes can be achieved (Haigh and Lanigan 1995).

The provenance of legislation from the EU has become much clearer in some areas than others. For example, environmental health officers have been very effective at understanding the system and becoming centrally involved in EU working parties and now with the new European Environment Agency. On the other hand, it has taken those involved with transport – the traffic engineers and transport planners – much longer to appreciate the power of EU policy in the UK. In the case of transport, the policy has been of a particularly broad sweep although it has had some very direct consequences such as in the slashing of much of the road programme and in the increased role of public transport. On the other hand, until recently, food regulation has been an area that civil servants have been happy to leave to technical staff in local government.

John Gummer, as Secretary of State for the Environment, broke with the obfuscation of the past in announcing publicly that over 80 per cent of UK environmental legislation is derived from Brussels and acknowledging the EU derivation of new legislation or initiatives such as the air quality plans and water quality improvement programmes. However, there is still some way to go to achieve a full degree of local government understanding, given the comparatively recent change in approach to transparency. In some areas, where EU policy is still at a formative stage, such as for the planning system, there remains on the whole a state of unpreparedness.

For local government in the 1990s, much of their funding is now achieved through partnerships, whether with organisations in other sectors or with local authorities in other parts of Europe. Increasingly, this is also the way they pursue their responsibilities for their localities more generally. On environmental issues, cross-sectoral partnerships have been focused mainly upon the local authority through the Local Agenda 21 roundtables. After the Rio Earth Summit, it was pledged that each local authority in the UK would be committed to and have taken some action on Local Agenda 21 by the end of 1996. In each local authority the specific priorities vary according to need and local conditions, but in many areas, the new round-tables have brought together social, economic and environmental partners for the first time (Morphet and Hams 1994).

The commitment to Local Agenda 21 represents a locality-based approach rather than one determined by specific services provided by the authority. It has also introduced a broader geographical perspective into the work of local government with increasing numbers of local authorities now linking with Third World countries. At the same time, the greater under-standing of the role of the EU in setting standards for environmental media has been achieved. In part this has been through the promulgation of the EU's Fifth Environmental Action Programme by the local authority associ-ations, together with local government's significant contribution to the review of the Programme (Morphet 1993). To this understanding have been added greater numbers of direct links with Brussels through cam-paigning networks such as Sustainable Cities, through funding schemes such as LIFE (see Note 2, p. 56) or THERMIE (a programme for the promotion of new energy technologies in Europe), or through direct influ-ence on EU-promoted projects such as the Trans European Networks (TENs).

INTERGOVERNMENTAL RELATIONS AND THE ENVIRONMENT

Much of the local authority lobbying of the European Union is undertaken behind the scenes through informal networks supported by representative offices in Brussels. Local authorities in Scotland have a joint representation as do those in Wales, but for England representation is fragmented into many small county-based offices. Such an approach does not compare well with that taken by other EU regions, which usually have unified offices often on quite a large scale, some exceeding thirty staff working full time in Brus-sels promoting their interests. These regional groups can also act together as country-based lobbies and it is significant that the first co-ordinated and regular meetings between all the UK local and regional offices and the Commission took place in early 1996 after an initiative by the English Regional Associations (ERAs).

Links are fostered with the other EU institutions, particularly the Parliament, and since 1994 local government has had its own institutionalised forum in the shape of the Committee of the Regions (COR), which was set up under the Maastricht Treaty. The Committee is composed of elected representatives of local and regional government, selected in the case of the UK by the Government but in the rest of the EU by the regional and local authorities themselves in joint discussion. COR has had its detractors, largely drawn from those whose influence would be reduced once it is fully operational. Its role has been contested, not least by some of the existing Union institutions, including the European Parliament and the Economic and Social Committee (ECOSOC). Initially it was seen to be a place where local politicians spent too much time discussing matters over which they had no influence. However, all talk of its demise as an institution has now ceased and it is expected to gain in status in the coming years (Morphet 1994).

Gradually the COR is establishing its presence. Its opinions are now tracked through the decision-making process of the Union and their influence is reported back to the COR. It is also having a major influence on the emerging approach to spatial planning which will be a significant determinant of the application of EU regional funds after 1999. Through a round of conferences, the COR has been a major factor in the development of mega-regional political structures which operate across member state borders. The COR is also seen as having a significantly enhanced role in the development of the EU. A growing number of member states wish to see a strengthened role for local and regional government at the heart of decision making. This is particularly important to totally federal states – Germany, Austria and Belgium – but also important to others, such as Spain, Italy and France, where regional powers are growing in importance (Morphet 1994).

The COR works through a series of Standing Committees (called Commissions), each chaired by a different member state (see Table 8.1). In addition there is a Bureau, or executive, made up of the leading members of each national group. The first President of the Committee of the Regions was French, the Vice-President Spanish and the Director-General German.

Since it commenced business, the COR Environment Committee has given opinions on a number of issues, including energy policy, the ecological quality of water, coastal management and environmental assessment. It also led the Opinion on the response to *Europe 2000+*, a major policy paper on spatial planning issued in 1995 by the Commission (CEC 1995c). In addition it made a major contribution to the environmental aspects of the Delors White Paper (CEC 1993c). The UK delegation at the COR has been well supported by the local authority associations and their advisers since the outset. In particular, the Environment Committee has been seen as an area of particular success for the UK, where there has been good co-operation

Table 8.1 Commissions (Standing Committees) of the Committee of the Regions (COR)

		Chaired by
Commission 1	Regional Development and Finance	Germany
Commission 2	Spatial Planning and Rural Policy	France
Commission 3	Transport	Spain
Commission 4	Urban Policies	Italy
Commission 5	Environment, Energy and Land-use Planning	Netherlands
Commission 6	Education and Training	UK
Commission 7	Citizens Europe, Culture, Youth and Consumers	Belgium
Commission 8	Economic and Social Cohesion, Social Policy and Public Health	Portugal

with the Dutch, Finns and French on issues such as energy policy, the environmental dimensions of trade and spatial planning.

The European Commission is also increasingly involving local authorities in the process of policy development in the early stages, in pilot projects, and through increased consultations with local authority networks. Although local authorities would argue that the Commission could be even more open to their influence, in reality it is possible to gain access more readily than would be the case in Whitehall, and Commission staff are generally more receptive. In part, this reflects the Commission's style of operation but also its recognition of local authorities as partners within the notion of subsidiarity. Local authorities are also seen to be a key contact with the citizens of Europe and the medium through which many policy changes can be brought into effect. However, much of the significant policy development in the EU is undertaken by member state governments under the auspices of the Council of Ministers and this often occurs without reference to the views of local authorities.

The process of developing spatial planning policy highlights the significant differences between the approach of the Commission and that of the Council of Ministers. The Commission started work on this in the late 1980s and published its key study *Europe 2000* in 1991 (CEC 1991c). It then developed further work undertaken by consultants and brought this together in *Europe 2000+*, published in 1995 (CEC 1995c). This report and its subsequent initiatives have been subject to a variety of inputs, have been discussed by COR, the European Parliament and local authorities, individually and collectively, and have been the subject of open seminars around the EU. In contrast, intergovernmental discussions on the future of planning policy have been undertaken behind closed doors by an informal Council on Spatial Development, made up of representatives from member state governments. This Council has been meeting since 1992. Its sessions are closed. Its work is undertaken by civil servants. It has published no papers. It is currently in the final stages of formulating a European Spatial Development Perspective introduced in 1997 under the Dutch Presidency. As this plan emerges, given its potential significance for the future pattern of regional funding and investment across Europe, it could receive an unsympathetic welcome. The failure to include local authorities at an early stage when they will be responsible for delivering much of it is likely to be a critical weakness.

NETWORKS

As part of its mission to stimulate greater European integration, the European Commission has specified that access to the vast majority of EU funding schemes depends on the formation of partnerships between organisations drawn from more than one member state. Over time, this has resulted in the establishment of networks through which local authorities are able to develop joint bids. These networks have often extended their scope, taking on a lobbying role. Some networks are based on their geographic proximity, such as the Atlantic Arc and the North Sea Commission. Others are based on a common interest – Eurocities, Car Free Cities, and Sustainable Towns and Cities being just three of the many established with the encouragement of the Commission to support the development and application of sustainable principles within the Union. Common interest networks have begun to prove to be an effective way of working together. They are able to lobby, exchange good practice and seek joint funding. For those authorities seeking partners, there is an annual meeting, Directoria, sponsored by the Commission, which serves as a marriage agency.

At a European level, local authorities join together in associations that are generally broader than the EU. The most active of these groupings are the Assembly of European Regions (AER), the Committee of European Municipalities and Regions (CEMR), which is the European Branch of the International Union of Local Authorities (IULA), and the Standing Conference on Local and Regional Authorities (CLARAE), which is part of the Council of Europe. These organisations work in much the same way as UK local authority associations, through standing committees. They are invited to comment on EU policies and proposals formally when these are published for consultation and have much informal contact. The Environment Dialogue Group of the CEMR and the Committee on Environment of the AER are both seen to be particularly important in establishing regular briefings and exchanges on policy. There is also the International Council of Local Environmental Initiatives (ICLEI), which is a local authority member organisation established after the Rio Earth Summit. Many UK local authorities have joined ICLEI and participate in its European and worldwide programmes, which have had some influence with European institutions.

The need for networking has been generated in part by Commission pressure for joint working and it has had some significant effects on UK local government practices. In the environmental area, there has been a considerable extension of understanding of what can be achieved within systems of local government through exchange of experience with other member states. Action on car-free cities has received a significant boost, for example, from practice in the Netherlands and Germany. In other areas such as Local Agenda 21, the influence of the UK has been stronger on other member states, with British traditions of voluntary action starting to influence practice in France, Germany and Denmark. Although at the outset, with networking, funding was the only issue of concern, concerted action to influence policy is now much more apparent. Whether this be on specific issues, such as coastal policy, or the review of the Fifth Environmental Action Programme, individual local authorities and the local authority associations have come to play a significant part in developing EU policy.

CENTRALISATION AND SUBSIDIARITY

The extensive Europeanisation of policy and the local authority response to it have coincided with major changes in the delivery of local services. Although many of the standard services are now no longer provided directly by the local authorities, but rather by the private sector or through arm's-length direct service organisations, there is still a responsibility placed on local authorities through a stream of legislation to provide services, in some cases to defined standards. It is this legislative approach, where each local authority activity is defined by statute, that marks out the UK from other European countries, where the functions of local authorities are enshrined in basic law. Thus, they are free to pursue actions which are appropriate for the well-being of their areas through the practical application of subsidiarity.

Haigh (1986) has argued that the increasing role of European legislation has led to a centralisation of environmental policy within the UK as central government is legally responsible for compliance within the Union and therefore has taken steps to control or supersede sub-central policy actors. However, this view fails to consider how the process worked before and how the legal and political framework of the EU offers scope for local authority initiative while constraining the actions of central government. In the past, for example, central government has always taken reserve powers or powers of last resort to intervene where it is deemed local authorities have not acted appropriately.

Although many environmental standards are set at the EU level, the local authorities now have a clear role in discussing their adoption and as ever have considerable freedom in determining how they should be met. So the new air quality measures should allow local authorities to assess their own problems and set their own programmes and actions to ensure that they meet the required standard. It is central government that has acted against the principle of subsidiarity which has made this ability to act fully at the local level more difficult.

When it comes to the national implementation of EU Directives, local authorities in the UK are often concerned and critical that their advice is only sought through consultation after the main points of the legislation have been agreed. This flies in the face of the shared responsibility principle and also fails to ensure that UK civil servants have the benefit of practitioners and local politicians to support their negotiating stance. Understandably, individual local authorities feel left in the dark, not just on specific issues but also on the legal and financial situations (Ward and Lowe 1994).

However, until the early 1990s, even the civil servants seemed not fully to appreciate the superiority of European law on matters such as waste, contracting procedures and water quality. European legislation was regarded as optional. However, various legal judgments have all affirmed the responsibility of UK courts to enforce EU law directly without the need for it to pass through Parliament.

This loss of sovereignty in a growing number of areas has been particularly difficult for civil servants to accept and has led them to adopt evasive strategies. For example, although the 1990 Environmental Protection Act implemented much of the Fourth Environmental Action Programme, it was dressed up as entirely domestic legislation. There has also been a strategy to maintain relations with sub-central actors as if central power was still in place – this is particularly so with the local authority associations which have been slow to realise the need to lobby in Brussels and not just in Whitehall and Westminster. In this policy community (Laffin 1986) the old boundaries have been maintained to perpetuate an inward-looking approach.

This loss of function has also caused civil servants to generate other strategies in relation to local authorities and in some cases to seek to take over local authority functions directly. This is the real source of centralisation. In particular the creation in the UK of the Government Offices for the Regions in 1993, established regional co-location for four departments – environment, transport, trade and industry, and employment – which now have their policy leads set within an EU competence. In establishing these offices, it can be argued that the civil service sought to undertake activities that were formerly undertaken at the local level, such as administering Structural Funds (Mawson 1995), as if by becoming the 'regional tier' they could preserve a role for themselves and some control. Graham Meadows, a senior member of the European Commission staff in DGXVI, has described this process as 'neo-colonialism'. All EU states, together with the aspirant members, now have regional structures or are putting them in place (Constitution Unit 1996). In order for localities to receive EU funding in future, they will need to have a regional, democratic tier of government. In Britain, the incoming Labour Government has begun to address this deficiency with its proposals for a Scottish Parliament, a Welsh Assembly and decentralisation through Regional Development Agencies in England, with the possibility of Regional Chambers or Assemblies to follow.

The attempt by the civil service to fill the regional vacuum will no doubt, in retrospect, be seen as a valiant exercise but one which was always inherently unstable and which did not resolve the strong departmental basis of government (Constitution Unit 1996). As indicated above, this compartmentalised approach in England has already been shown to be less effective than a unified approach in Scotland and Wales. The Government Offices have failed to resolve many of these issues (Mawson and Spencer 1997), while increasing the extent to which they have exercised power over funds without explicit ministerial consent (Constitution Unit 1996).

At the same time, the style of policy making in Brussels has also been shifting. Under Delors, the Commission was run on competitive lines, much the same as the UK civil service, which reinforced a departmentalised approach. However, under the new European Commission President Jacques Santer, a real change is emerging towards an *integrated* and *programmed* approach. Thus recent reviews of transport policy (CEC 1995f), environment policy (CEC 1996d) and spatial planning policy (CEC 1995c) are all overlapping and interconnected. At the same time, they will all have profound implications at the local level. For example, the promotion of the proximity principle in terms of environment, transport and energy could lead to a tax on the movement of goods. For low value goods, this might encourage a closer relationship between rural hinterlands and their urban hearts, with implications for the distributive, retail and agricultural industries. Will the new regional plans be required to address such structural change?

More critically, these policy reviews, together with what is expected to emerge from the review of the CAP, are all relying on the spatial planning system to deliver at the local and regional levels. At the same time, the Commission is encouraging the development of groups of regions in new working arrangements. These mega-regions, identified in *Europe 2000+*, are not groupings of member states but new geopolitical alliances – the Atlantic Arc, the North Sea, and the Centre and Capitals regions, for example. These will be required to develop new planning frameworks to deliver EU policy and possibly funding at the sub-national level. The model approach is the Baltic Sea Plan (Baltic Institute 1994).

Although the Commission and member state initiatives are developing side by side, it is possible to interleave their approaches as set out in Figure 8.1.

Mega-region

/

Region

/

Integrated programmes

urban (former Structural Funds 1 and 2)

rural (CAP and 5a, 5b and 6)

coastal

mountains

extra-urban

/

Integrated (unified) system of development permits

(to meet the requirements of the Single Market)

Figure 8.1 European spatial planning: a possible framework

If this approach does emerge, it will place local and regional authorities in a strong and central position, not least as they will be responsible for the development of policy below the EU level. At the same time, the Committee of the Regions is making its case to be centrally involved in the development of European spatial planning.

All this leaves the future of central–local relations in the UK in a position of some uncertainty. Within the UK, the review of local government has largely distracted local authorities from these broader issues, including the need for a democratic tier of regional government within which to focus locality based demands. If the discussions on constitutional reform initiated by the Labour Party in opposition come to fruition then indirectly elected regional chambers should translate into directly elected regional assemblies, leading to further change in central–local relations (Straw 1995). There will still be a need for a government presence in the region but it seems unlikely that the Government Offices could remain in the current format as many of the issues under their control would be transferred to the Assemblies, albeit without necessarily transferring central government staff in the process. The new Assemblies will have the EU interface as a prime concern and at this point the use of the spatial planning system to deliver improved environmental standards within the regional plan will become apparent.

Where do these developments leave Haigh's argument (1986; Haigh and Lanigan 1995) that the increased role of the EU in the development of policy and standards has left the local authorities more powerless than before? It would seem instead that EU pressures have forced the civil service to inhabit a lower tier of government in the environmental arena through the creation of the Government Offices. The consequences of the establishment of the UK Environment Agency are still unclear but they are as likely to lead to a loss of central departmental policy initiative as to any loss of local regulatory responsibilities.

Haigh's view may represent an interpretation of the position at the time of writing, but the changes in approach which have occurred since the Rio Earth Summit and the adoption of the Fifth Environmental Action Programme suggest that local government is having influence and that it is to be the main agent of delivery. It is central government, in its isolationist approach to Brussels negotiations (in this case, by not working with domestic partners) which has lost power. Under the current system of management of business, Parliament is also unable to inject its views until after EU legislation is agreed. The reform of the role of domestic parliaments *vis-à-vis* EU legislation, as proposed in the UK's White Paper on the Inter Governmental Conference, reaffirms this need (UK Government 1996).

Despite the major changes in the operation of local government over the past twenty years in the UK, the provision of environmental services remains a central function for local authorities. These services are usually for the benefit of all members of the community, unlike, say, education or social services, which are client-group directed. Over the past ten years, the public has come to view the local authority as the foremost environmental guardian in its locality. This role has often been too difficult to maintain since authorities have needed to respond to growing requirements for housing land and out-of-town development as government policies and directions have changed (Morphet 1995). However, it is also the case that through activities such as recycling, environmental charters, Local Agenda 21 and environmental management systems, local authorities have come to be seen as leaders and facilitators in the community (Morphet *et al.* 1994).

THE IMPACT AND INFLUENCE OF LOCAL AUTHORITIES IN THE EUROPEAN ENVIRONMENTAL DEBATE

In terms of influencing the development of policy and legislation, local authorities have always worked through their local authority associations. It was during the lobbying for the 1990 Environmental Protection Act that the local authorities began to realise more clearly that the UK parliamentary machine had less influence over the content of legislation than it once had had.

The awakening consciousness of local government on this influence ran in parallel to new considerations of the approach to environmental initiatives in Brussels. The Fifth Environmental Action Programme was based on the principle of shared responsibility, which is stated from the outset but it is also underlined in two distinctive ways. The first is through the seventeen sectoral and thematic programmes of action, where the responsibility for implementing specific actions is identified: in many cases this is local government. The second is through the delineation of shared responsibility, where the local authority role is again clearly identified.

When the Fifth Action Programme was first published, it received little publicity. Indeed not many people understood its purpose and role. In the UK, at best it was seen by government as a Commission 'wish list' or merely a collection of initiatives already in the pipeline and easy to implement. However, the UK local authority associations considered it of sufficient importance to their membership to commission a guide to the Programme that would cut through the Euro-language to reveal clearly the commitment to the local government role in the delivery of the environmental agenda over a ten-year period.

However, even where local authorities were identified as having a role there was no real attempt in the UK to bring them more centrally into the policy process. Nevertheless, local authorities were responding enthusiastically to the Rio challenge, employing Local Agenda 21 co-ordinators and undertaking other key activities such as the production of environmental audits and indicators (Ward and Lowe 1994). They were also keen to take up the challenge of the Fifth Environmental Action Programme.

The local authority associations therefore sought to gain more influence on the policy agenda. Within Whitehall, this was through existing networks. These were already well established. The UK, for example, was one of the very few states to have included local authority representatives as part of its delegation to the Rio Earth Summit. The major success in this approach has been through the Central and Local Government Environment Forum, which is cross-sectoral and meets regularly. However, local authorities have been kept to a minimal presence on the UK Roundtable on Sustainable Development set up by the Prime Minister in 1994. Finally, some influence is exerted through the local authority professional associations as well as through professional bodies with strong local authority links such as the Chartered Institute of Environmental Health.

The growing influence of UK local government within the Brussels machine is demonstrated by the approach to the Fifth Environmental Action Programme. When it was being prepared prior to adoption in 1992, there was no involvement of UK local authority interests, but four years later, they were actively involved in the review process. The outcome of the review, published in January 1996, accorded, in its action plan for the next five years, an even more central role to local and regional authorities in the development and implementation of policy (CEC 1996a). Although there are no explicit links to Chapter 28 of the Earth Charter, as UK local government would have wished, there is an overall shift towards Rio in the review. This approach was much supported by UK local government in its submission to the Commission (LGMB 1996). The submission reinforced many of the concerns that had been at the forefront of the regular meetings between UK local government and the Commission, and it is clear that many of them have been incorporated into the review.

In some areas, UK local government is acknowledged to have taken the lead in the implementation of the Fifth Environmental Action Programme. These are the Eco-management and Audit Scheme (EMAS); sustainable indicators; and incorporating environmental requirements into competitive tendering. Through the joint development of the local government EMAS scheme, UK local authorities have been able to demonstrate both their commitment to practical environmental matters and support for EU initiatives. Although similar initiatives such as the Packaging Directive and eco-labelling have taken longer to process, EMAS has found considerable support across UK local government. Based on the EU's industry scheme, the local government EMAS has also demonstrated the ability of the scheme, which was process-directed, to be adapted to the service sector. Also through its work on sustainable indicators, UK local government has contributed to the development of European policy in the field of environmental information and monitoring, a field which is otherwise largely the preserve of central governments through the European Environment Agency (see Chapter 7). Finally, in seeking to ensure that environmental objectives were not lost when services were privatised, the Local Government Management Board, acting on behalf of the local authority associations, was able to derive approaches with the DoE to ensure that these objectives were maintained in contract specifications and evaluation, based on the European rules set out in Directive 92/50.

The international co-operation element of the Fifth Environmental Action Programme was considered by UK local government to be particularly weak. However, the reference to Agenda 21 in the review suggests that the broader concepts adopted in Rio are seen as central to the next five years. UK local government particularly wished the Commission to make a submission to the 1997 meeting of the UN Commission on Sustainable Development on the implementation of Agenda 21 in Europe. It has also been urging that local government's role in the delivery of Agenda 21 would be specifically acknowledged.

At this early stage, it would be possible to see the review of the Fifth Environmental Action Programme as a major success for local government. There are aspects of Agenda 21, including links with equity, poverty and health issues that have still not been explicitly embraced. There is no mention of support for an international environmental court and no mention of the role of the various institutions of the EU (Council of Ministers, Committee of the Regions, and so on) in achieving sustainable development.

As the UK Government assessment of the Fifth Environmental Action Programme has demonstrated (DoE 1994), all commitments have been implemented in some form and the majority in the way outlined. The review proposals are likely to be far more radical than those in the 1992–7 version as they will be linked more centrally to the machinery of state such as taxation, trade policy, national accounts and risk management.

Although understandings about the interrelationship between local government and environmental policy vary between local authorities according to their own interests and priorities, where there are major environmental issues to be faced the need for action in Brussels is well appreciated, whether this be on water quality, air quality, energy or transport. Understanding is therefore primarily on an issue basis. Where there is often little understanding is at a more complex and integrated level, for example the spatial implications of the application of EU policy. However, this is an issue still very much confined to discussions between the Commission and member state governments. As this chapter explains, this position is expected to change shortly.

For the most part, also, these matters are more the concerns of local authority officers rather than of the councillors. Nevertheless, there are some members, such as Councillor Sir John Harman of Kirklees or Councillor Peter Soulsby of Leicester, who have championed the need for an environmentally led integrated approach within the local authority associations and regional bodies. However, there is less leadership on these issues at the national level in any of the parties, with the Liberal Democrats perhaps offering more understanding and leadership than the other two main parties. The extent of European environmental legislation yet to come in the next five years could be a more significant determinant of policy development than the colour of the government.

Until recently, a similar comment could have been made about the role of Europe and its effect on local government. However, during 1996–7, the level of understanding about the scope of influence and potential for change dramatically increased as a part of the constitutional debate preceding the 1997 Treaty revision.

CONCLUSIONS

The major barriers for local authorities in coming to terms with the role of the EU in determining policy and legislation have been:

- levels of understanding in local government;
- the negative outlook of central government towards European and environmental matters;
- the nature of policy communities within the Whitehall system which have incorporated the local authority associations and central departments in an inward-looking relationship in which neither has wanted to recognise its dwindling influence;

- the system of government in the UK in which local authorities are creatures of statute rather than possessing general powers of competence.

However, over the period since 1992, UK local government has developed a significant understanding of the way in which it can gain influence in Brussels and has mobilised in order to achieve this (see Chapter 5). This has primarily been through policy development and in some ways is completely at odds with the more common ways of local authority working with the Commission. The approach has started to bear fruit, and credibility has been established. This always has to be tempered by the stance taken by the UK Government in all European negotiations. Much of the work of the future needs to concentrate on gaining involvement within discussions of European policy from the UK end. At present this door is entirely closed and thus the fine tuning of European environmental policy to UK needs is inevitably poorer as a result.

The future offers more fundamental opportunities. The central incorporation of the role of local government as one of the main agents of achieving the European environmental agenda could provide local government in the UK with the boost it is seeking. The advent of a new government more sympathetic to local governments as well as to the EU and the environment provides a much more auspicious context.

Local practice, however, will continue to adjust to changing local, European and global pressures. The requirement for Air Quality Plans (arising from a commitment in the Fifth Environmental Action Programme) demonstrates the influence such requirements will have on many issues at the local level, such as planning, traffic management, public transport investment, and school travel arrangements. Each local authority area will concentrate on those matters which are more relevant to it. Air quality is more important in Greenwich, water supply in Yorkshire and minerals in the Mendips. The understanding of the overarching role of the EU in setting environmental standards is now growing in the public consciousness and this will inevitably have considerable spin-off at the local level.

9

BUSINESS LOBBYING ON THE ENVIRONMENT

The perspective of the water sector

Edwin Thairs

Today monitoring of EU developments is an established fact of life for British business. The Single European Market poses tremendous opportunities (and a few threats). To operate in and from it, in competition with other European companies, British business wants to be assured of a level playing-field, not least in relation to environmental regulations. In many cases major investment and operational decisions at home are also greatly influenced by regulations initiated in Brussels.

This chapter examines the development of environmental policy making at the European level and the organised response to this of the business community. It is written from an insider's perspective by someone who has long been involved in advising British business and industry on European matters. In recent years this experience has been specifically with the water sector, and the chapter is illustrated with examples of the sector's changing involvement in European environmental policy.

European regulations have enormous implications for the water industry. From 1989 to 1995, the water and sewage companies of England and Wales invested some £15 billion, mainly on sewage treatment, sewerage and water quality. A further £14 billion of capital spending is planned until the end of the century and another £10 billion from 2000 to 2005. Since legislation is the principal driver of these investments and most environmental legislation now stems from Europe, the European Union is responsible for the bulk of these costs, or about £1–2 billion per annum.

There is another reason for the UK water industry to be interested in the EU, namely in relation to its non-core business activities. The European market for environmental equipment was estimated to be about £75 billion in 1992, of which a quarter was in water and waste water. By 2000, the market should be £125 billion, with roughly the same proportion attributable to water services. As these figures demonstrate, water pollution control is fast becoming a significant business in its own right, prompted extensively by EU legislation.

BUSINESS AND THE EUROPEAN ENVIRONMENTAL ARENA IN THE 1970s

The early political organisation of business at the European level

Business involvement in Europe has not always been so extensive. When the UK joined the European Economic Community (EEC), British industry was primarily concerned with what the commercial opportunities might be. The Confederation of British Industry (CBI), which had supported British membership, set up an office in Brussels to monitor developments and to provide a focal point for enquiries on European matters. It also promoted the development of the Union of Industrial Confederations of Europe (UNICE), a European association for national bureaux of industry. At this time, few individual sectors of business were present. None the less, through the CBI and UNICE, UK industries began to assume an active role in European environmental matters, recognising from experience of their international operations and trade that transnational environmental issues were of growing importance to the well-being of their businesses. UNICE operated via a committee structure, with each national confederation of industry entitled to appoint two members and each relevant European sectoral trade association an observer. The CBI nominated the UK members to UNICE's Environment Committee, normally one from its permanent staff in London and another from a company with European interests. UNICE's permanent secretariat had limited resources, with, for example, the Industrial Development Manager responsible not only for all environmental issues but for energy, transport and technology too. This state of affairs has changed little over the years, with just one person in UNICE even now working less than half time on environmental matters. In the early years, though, the demands were not onerous, given the relatively small amount of environmental legislation emanating from Europe.

In addition to these broad business and industry routes into Europe, in 1976 the water sector (through the then regional water authorities) collaborated in the creation of an organisational structure specifically for the European water industry known as EUREAU. Essentially, EUREAU acts as a confederation of associations of water suppliers in all countries of the EU and the European Free Trade Association (EFTA). Its initial function was to provide a forum for the exchange of technical information and practical experience of operating water supply and treatment systems. This led to a number of operating guidelines and standards for the sector, ensuring appropriate levels of treatment and distribution of drinking water throughout Europe. It was only in the 1980s that seeking to influence European legislation and its quality standards assumed an equal status. The EUREAU Board of Management is now served by three Committees respectively on water quality and resources, technical standards, and economic and legal aspects, each with representation from all member countries. These contribute operational expertise to the papers and reports that EUREAU produces, mainly as the basis for detailed discussions with the European Commission.

European environmental policy formation in the early years

In 1972 the European Commission had established an Environment and Consumer Service with the initial remit to prepare a Community Action Programme on the Environment. The Service was a minor part of the Commission and its Head did not enjoy the status of a Director-General. Access to it was somewhat limited, and industry was not geared up to liaise with it. The main access was through national Ministries and the European institutions, notably at that time the Economic and Social Committee, representing employers, employees and consumers. For any relevant proposal from the Commission, including those on the environment, a small task force was established, charged with drafting the Committee's Opinion: to do this, external experts were brought in, the employers' expert usually being chosen in consultation with UNICE or the sectoral European trade association with the greatest interest in the proposal. The Committee could also make 'own initiative' reports on any topic which it considered the European Community should be addressing, for example on the relationship between the environment and jobs.

The lack of business organisation and influence over early European environmental initiatives is well illustrated by the abstract and over-ambitious nature of the First Environmental Action Programme issued in 1973, especially its precise but quite unrealistic timetables. None the less the principles (set out below) in the First Environmental Action Programme were important and set the policy framework for what was to come:

- prevent pollution or nuisance at source;
- take early account of the environment in decisions;
- avoid overexploitation of resources and nature;
- improve scientific and technological knowledge;
- the polluter should pay;
- avoid adverse transboundary effects;
- take account of the interests of developing countries;
- promote global research and policy;

- environmental protection should concern all;
- take action at the appropriate level;
- integrate policies and action.

In practical terms, the early years of the Community's environmental programme were characterised by the search for legislative controls over specific problem areas, such as the 1976 Dangerous Substances Directive, the 1978 Bathing Water Directive and the 1980 Drinking Water Directive.

Dangerous substances discharged to water

The Community's effort to regulate discharges of dangerous substances to the aquatic environment caused particular difficulty for the UK, challenging its traditional approach to environmental controls. Hitherto the UK had sought to maintain the quality of receiving waters in relation to the uses to which they were put by imposing conditions locally on consents to discharge (see Chapters 10 and 14). The German Government led the call for uniform discharge limits, pointing out the administrative efficiency of regulating all discharges to the same standard irrespective of where the discharge was made and the uses to which the receiving water was put. UK industry, as represented by the CBI, argued strongly through the Department of the Environment against the unfairness and practical consequences of dischargers being subject to fixed emission limits irrespective of where they were located (see Chapter 2). Many companies had located on estuaries or near the sea partly because of the natural dispersion and dilution these waters had to offer. In the end both positions were accommodated: the UK was allowed to continue with its environmental quality approach for a defined list of the more dangerous substances; other countries could opt for fixed emission limits. Other dangerous substances covered by the Directive would be controlled by national programmes based on environmental quality criteria. Maybe the most significant impact on the UK was the introduction of the term 'environmental quality objectives' and the formalisation of the system through the establishment of specific quality standards.

Bathing and drinking waters

Although the Dangerous Substances Directive involved the CBI preparing position papers for officials and ministers, much of the early legislation involved little technical input from industry. It was expected that the British Government would look after the interests of British industry in Brussels negotiations and that, in any case, the details could be ironed out in the subsequent implementation of Directives within the UK.

A clear example is the Bathing Water Directive of 1976. The chemical and manufacturing sectors saw little in the proposal to concern them. Indeed there was a general presumption in the UK that the Directive would be most relevant to Southern Europe where tourist populations and interest in high-quality bathing waters would be pronounced. The Government therefore negotiated the Directive with little reference to or assistance from industry in general or the water sector in particular. It did, however, negotiate a text which allowed for a fairly limited implementation within the UK.

During its initial years, the Department of the Environment was also well endowed with scientists who helped to ensure a sound scientific basis for national policies and whose expertise shaped the Government's response to proposed Directives, such as the 1980 Drinking Water Directive. However, the switch in environmental initiative from Whitehall to Brussels set policy staff and administrators into the ascendancy in the Department, leading to the gradual decline of its internal scientific capacity, and bringing it more into line with the Commission's own staffing arrangements. In the 1970s, though, the Department of the Environment was still well served with scientists, and on questions of water regulation it could call additionally upon the advice of operators via the National Water Council. The technical input into policy development tended to be through established networks: written technical submissions at that time are conspicuous by their absence. Such arrangements, however, helped to achieve, in the Drinking Water Directive, a piece of legislation which, although elaborate and over-prescriptive, provided a benchmark of quality for all drinking waters in the Community.

BUSINESS AND THE EUROPEAN ENVIRONMENTAL ARENA IN THE 1980s

Increased business representation in Brussels

By the early 1980s, more than half of UK trade was with other European countries and there was a build-up of a substantial presence of British business representation in Brussels to reflect this. Several companies and national associations felt that they needed their own European office. Many more employed consultancies out of a need for regular intelligence from and contacts with European bodies, but not to the extent to justify the expense of setting up and running a separate office in Brussels. European political monitoring consultancies were often an extension of the services provided nationally by parliamentary or public affairs consultancies. For a fee, the consultancy would monitor political developments in the European institutions of interest to its clients, advise on the most appropriate course of lobbying or other action, and ensure that the clients' views were heard by those who took political decisions.

European sectoral trade associations also became more active. For example, those for the chemicals industry and non-ferrous metals industry organised annual meetings with relevant staff from the Commission to consider all environmental issues of current concern to them. These meetings took place in different member states, usually in association with a visit to an industrial plant in the area. Other sectoral associations prepared detailed reports for the Commission and other policy makers in Europe: for example, the oil industry prepared technical reports on aspects of its operations which impacted on the environment, such as the sulphur content of fuel oil. Such ways of informing the Commission and policy makers laid the foundation for more overt political lobbying.

In 1983 the regional water authorities in England and Wales became freed from their local authority masters, although remaining in the public sector. The somewhat bureaucratic National Water Council was abolished, to be replaced by a trade association – the Water Authorities Association – which quickly began to promote independent advice on a range of issues, including new EC legislative proposals. Then, in the late 1980s, privatisation took the water services companies out of the Public Sector Borrowing Requirement, enabling them to compete for investment funding on the open market. It also liberated them from the Department of the Environment allowing them to make direct contact with the European Commission and other institutions. The Water Authorities Association was reformed as the Water Services Association and given a specific function of looking after the industry's collective interests in Europe – a role made easier by the early appointment of a consultancy in Brussels to monitor relevant developments. The privatised companies were also able to join the CBI, which gave them another channel of access into both the British Government and the European Union.

European environmental policies in the 1980s

The European political climate changed dramatically in the early 1980s. The German Greens made huge strides and the German Government, which hitherto had often sided with the UK in being cautious and reactive to European legislative proposals, suddenly became their leading promoter. The European Parliament, subjected to direct elections for the first time in 1979, started to show its teeth.

The need for unanimity for all environmental proposals disappeared in the Single European Act of 1986, although remaining for those solely of an environmental nature. The UK no longer always had the last resort of saying 'no' if it was outvoted by its European partners – so informal discussion, persuasion, and reliance on sound scientific data became more central features of its negotiating tactics. The Government – and the Department of Trade and Industry in particular – increasingly looked to industry for the data and examples on which to base its arguments in Brussels. Although the opportunities for industry to influence decision making were broadened, these opportunities were not without problems or risks. The relevant information was not always easy to come by, including often the latest text of a proposed Directive. Impact assessment techniques were undeveloped and, where estimates and projections had to be made, British industry ran the risk of being accused of exaggerating the costs and practical consequences. Comparative information on the costs to industry in other member states was often lacking. British companies with subsidiaries elsewhere in Europe were reluctant to use internal information where this would draw attention to the favourable conditions (such as subsidies) under which these subsidiaries operated.

Business representations helped to provide the impetus to complete the European market which the Single European Act boosted by removing internal barriers to trade for a range of products and services. At the same time, the Act gave European environmental policy a legitimacy, introducing environmental provisions into the Treaty for the first time (see Chapter 1). It confirmed the guiding principles as

- preventive action;
- rectification of environmental damage at source;
- the polluter pays.

Although not using the term, the Act also established the principle of subsidiarity in relation to European environmental policy – EC action would be taken where objectives were better attained at the Community level but otherwise action would be taken nationally or locally. The principle should not prevent member states taking or maintaining more stringent measures than those decided at Community level. Industry supported the idea of action being taken at the most appropriate level so that local environmental situations could be taken into account. It saw EC involvement as ensuring that a basic framework of control was in place to protect the environment, leaving member states to work out the detail and to report back to the Commission on implementation and performance. However, industry also felt that the EC should lead where, for environmental reasons, product standards needed to be harmonised or marketing and use controls introduced. The EC should also have a key role in achieving consistency in sampling and analytical techniques, in assessing and reporting compliance; and in initiating enforcement actions if member states were unduly lax in transposing agreed Directives.

BUSINESS AND THE EUROPEAN
ENVIRONMENTAL ARENA IN THE 1990s

By the late 1980s the environment had risen to the top of the popular agenda, giving renewed impetus to the proposals and policies which the Commission were initiating. Many of the issues went beyond the boundaries of the European Community and were developed in international fora, with the Commission becoming the agent for translating the results into European law (Haigh 1992).

The significance of EC regulations was now fully recognised by industry across Europe. The breadth and detail of their coverage required ever more focused technical inputs from across European industry. This led to the establishment of a growing number of specialised industrial lobbying associations organised on a European basis. The formation of the European Waste Waters Group in response to the Urban Waste Water Treatment Directive exemplifies this trend. The efforts of British and European policy makers to integrate environmental considerations fully into business planning and management – through the promotion of such concepts as integrated pollution control, environmental auditing and sustainable development – necessitated closer consultation with industrial interests more generally.

The Urban Waste Water Treatment Directive

Agreement to end the marine deposition of sewage sludge was reached at a North Sea Ministerial Conference and implemented through the Urban Waste Water Treatment Directive of 1991 (see Chapters 2 and 14). This Directive was an essentially prescriptive piece of regulation setting down levels of sewage collection and treatment that were to demand substantial investment throughout Europe (see Table 9.1). Although the Directive places major demands on the UK waste water sector, the need for it and the

Table 9.1 Estimated costs of implementing the Urban Waste Water Directive in selected European countries

	billion Ecu
France	20.0
Germany	160.0
Ireland	1.4
Italy	10.0
Portugal	1.2
Spain	9.8
UK	11.0

principles underlying it were not challenged. The sector accepted the case for a Directive specially focused on its activities: it was discharging far more than any other sector to watercourses and there did need to be consistent levels of sanitation for EU countries. It was important, however, that the Directive was practicable: the waste water sector has very limited opportunity – and often none at all – of discontinuing operations to sort out technical problems of compliance. Further, the implementation of the Directive needed to be phased in such a way that financing of new borrowing and additional charges to customers could be planned and arranged to avoid steep increases in bills.

The British water industry was reasonably well placed to implement the Directive, starting from a position where 96 per cent of the population were connected to sewers, with 83 per cent of sewage receiving at least primary treatment. None the less, the Water Services Association estimated in 1991 that the Directive should still cost the UK £6 billion or more in capital expenditure, mainly to put in place new collection and treatment facilities around the coast where hitherto the policy of 'dilute and disperse' had been practised, but also to incorporate full secondary treatment at all existing inland sewage treatment works and nutrient removal when they discharged to sensitive areas. This estimate was borne out by later studies which show the cost of implementing the Directive at £6.1 billion in England and Wales and £8.7 billion in the UK as a whole. With such large sums involved, it was important for the text of the Directive and the subsequent implementation programme to be as practicable and efficient as possible. The Water Services Association and its members, the newly privatised water companies, closely monitored and commented upon the practical details of the proposed Directive. Having only recently come into existence, they were not in a position to negotiate directly with the Commission or other European institutions, and most of the lobbying took place through the Department of the Environment.

The emergence of specialised European industrial lobbying associations: the example of the European Waste Waters Group

The Urban Waste Water Treatment Directive demonstrated that the waste water sector was not sufficiently co-ordinated across Europe to present a united view at Community level. In principle, EUREAU could have taken on this role but its membership was predominantly water suppliers, whereas waste water collection and treatment usually remained the responsibility of separate public authorities. In several parts of Southern Europe, facilities for waste water collection and treatment were poor and sometimes non-existent. Even where they existed, their administration was very localised, with little contact even with neighbouring authorities, let alone co-ordination at national or European levels. The European waste water sector had no immediate means of uniting to present its practical experience to those preparing the Directive.

Waste water operators in France, Spain, Portugal, Italy and the United Kingdom quickly realised that, faced with the Directive, they needed to get together to exchange experiences and views, and to plan a more consistent approach to European developments affecting their common interest. A loose amalgamation of associations representing waste water operators in these countries was forged. This became formalised into a new association, the European Waste Waters Group (EWWG). The five founding members were quickly joined by other member states, so that the Group became truly representative of EU waste water operators.

The management style of the European Waste Waters Group was entirely different from that of EUREAU. A President was elected (the first being the then Chairman of Severn Trent Water) and he provided the Secretariat (initially the Water Services Association), both for a period of two years. Initially all discussions and decisions were taken in a plenary session of members, although subsequently this was reshaped into a Board and a Technical and Economic Committee (comprising one person, plus a named alternative, from each country) formed to manage and take action on detailed issues. Flexibility was the essential feature of the Group's operations, allowing it to respond rapidly to EU proposals, without elaborate approval procedures for the positions it proposed to take. By contrast, EUREAU's positions were worked out by committees and needed Board approval, procedures which could take up considerable time and effort. Streamlining of procedures and administration in recent years has paved the way for a merger between EUREAU and the European Waste Waters Group, taking effect at the beginning of 1998.

The initial and continuing task of the European Waste Waters Group was to remind the Commission and Parliament that a consequence of the Urban Waste Water Treatment Directive is the generation of more sludge, for which land-based outlets have to be found. In 1994, over 250 million tonnes of wet sewage sludge was generated in the European Union (equivalent to 12 million tonnes of dry solids, the usual way of measurement). This will increase by up to 30 per cent as more sewage is collected and treated. At the same time, deposition at sea must end in 1998, and proposed Directives on landfill and on incineration will further restrict the availability of these routes of disposal. However, sewage sludge is an inevitable and increasing consequence of treating waste waters, and European recycling and waste strategies must be framed to accommodate it (see Chapter 11). In short, water protection policies must be properly integrated with waste management, the regulation of air pollution and other environmental policies if they are really to be effective.

Integrating environmental considerations into business planning and management

The theme of integrated pollution control, or holistic environmental management, is, of course, one of the characteristics of the 1990s. Here, a UK initiative influenced the introduction of Community legislation, although the notion of overall regulation of all discharges from an industrial site is not new and has been successfully applied in a number of countries. The CBI supported its introduction in the UK, under the Environmental Protection Act 1990, hoping for a one-stop-shop Inspectorate able to balance what it allowed industry to discharge to air, land or water to achieve the best practicable environmental effect. Industry also hoped for practical guidance on how it should meet the required environmental targets (see Chapter 15). These developments were a precursor to the Community's Integrated Pollution Prevention and Control (IPPC) Directive.

This example illustrates the changing attitude of the UK to EC environmental policy, from a defensive position in the early years to providing some leadership and proactivity in recent years, a shift that UK industry has broadly welcomed. Another example is the Eco-Management and Audit Scheme (EMAS), which was also influenced by an initiative taken in the UK, this time by the British Standards Institution in its environmental management standard BS7750. Clearly the BSI takes forward the interests of the business community, and so does the Community legislation in introducing EMAS.

Another key theme of the 1990s is sustainable development. Although difficult to operationalise, in practical terms it should involve a careful assessment of actual and potential impacts on the environment of business processes and products (including life cycle analysis), with adjustments then made to curb these impacts. The consequent steps – of minimising waste generation, conserving natural resources, energy conservation and efficiency, using water wisely, sound transport policies, maintaining biodiversity, and so on – command the support of industry in principle as the sorts of actions that increase business efficiency, provide new market opportunities and strengthen public and customer credibility.

These concepts were elaborated in the Fifth Environmental Action Programme of 1992. Because successive Action Programmes have set the medium-term agenda, British business has always taken them seriously and sought to influence their content and philosophy. The proposed Fifth Environmental Action Programme was no exception, attracting comment from the CBI and the Water Services Association (as well as from other national and European trade associations). These comments were submitted to the European institutions, the government Departments responsible for negotiating the final text, and the House of Lords Environment Sub-Committee which conducted its own thorough analysis of the Commission's proposals. Broadly speaking, British industry supported the framework that the Commission suggested:

- the importance of all sectors of society being involved in improving environmental quality, and encouraging partnerships between them;
- selecting targets (industry, transport, energy, agriculture, tourism) for priority attention;
- improvements to environmental data; research and development; education and training;
- use of a wider range of policy instruments;
- more EU involvement in global environmental initiatives.

Consultation with industry

The Fifth Environmental Action Programme foresaw good consultation as a prerequisite for getting practicable and enforceable regulation and commitment to new environmental protection measures, be they legislative or voluntary. The practice of using Green (and White) Papers to sound out opinion became more prevalent, for example in relation to environmental liability. Formal arrangements were established for discussing major items with business leaders and other interests whose commitment would be crucial to the success of EU environmental policies. These were Commission initiatives, for which the business community itself had not been particularly pressing, and which caused some initial concern to European trade associations such as UNICE, EUREAU and the European Waste Waters Group who provide a more democratically derived viewpoint from the industries they represent. However, with the assurance that the new consultative fora set up by the Commission would not bypass the more representative industrial groups, industry saw merit in another channel of communication with the Commission through its business leaders.

Consultative conferences have been organised by the Commission and the Parliament involving experts from the business community and other parts of society. A notable example for the water industry was the 1993 workshop on revising the Drinking Water Directive. This gave all those with interests in the Directive – environmental, agricultural, general industrial, and political, as well as the water suppliers – an opportunity to give their views on how, if at all, the 1980 Directive should be changed. The Commission was able to gain from the workshop an impression of the balance of opinion, hence helping it to decide how far to go in proposing amendments. Other participants were able to gauge the level of support for their views, hence helping them to decide further lobbying activity. It is too early to judge how influential the organised business input to these consultations has been – but the water sector is optimistic that the mutually compatible goals of environmental care and business development can be taken forward in a practical way.

The basis for gauging environmental improvement and business performance must, of course, be adequate and reliable data. Business has provided the tools to enable more and better data to be collected, analysed and presented. The Commission and, since 1993, the European Environment Agency, have been making great strides in using these tools to pool, compare and disseminate the data. The overall result will be a much better knowledge base on which the European Union and member states can take policy decisions and for highlighting the responsibilities of all sectors of society towards environmental protection.

We have now reached a 'review' stage in the development of European environmental policy, involving, among others, an update of the waste strategy adopted in 1989; a review of the Fifth Environmental Action Programme; a review of its water policy and the development of a framework Directive; an assessment of results of the debate on environmental liability; and the development of a groundwater action programme. The business community has made its views known on all of these.

BUSINESS LOBBYING TACTICS: THE EXAMPLE OF THE WATER SECTOR

The water industry sees its main role in the development of European environmental policy as providing factual information and opinion on the practicability of proposed standards and other possible requirements that affect its interests. With its managerial and operational expertise, the UK water industry is well placed to answer the Commission's more technical enquiries. Much of this is arranged through EUREAU and the European Waste Waters Group, which can draw together data and experiences from operators in several member states. Within the UK, water and sewage companies are required to provide substantial information to the Drinking Water Inspectorate, the Environment Agency and government Departments, for example to allow monitoring of compliance with statutory standards. In addition, the water industry puts a great deal of information voluntarily into the public domain. Most of these data are also accessible to the Commission via the published reports of Government and the industry's regulators or through public registers or voluntary submissions. The water industry is ready to point out the practical and economic consequences of the Commission's legislative proposals, so that these can be framed to be operationally efficient and environmentally cost-effective.

In exchange, the water industry expects to be fully consulted via its European trade associations – EUREAU and the European Waste Waters Group – on any proposals likely to affect its business activities; for final standards to be sensible and uniformly implemented and enforced throughout the Union; and for the framework of control to be stable, with any changes signalled well in advance so that investments can be properly planned.

The main lobbying target for the British business community in general and the water industry in particular continues to be the UK Government, even on European matters. In the end, it is national governments that negotiate programmes, plans and proposals within the Council of Ministers. The Department of the Environment is the main government department for the water industry with overall responsibility for water policy and environmental protection. The Water Services Association also maintains contact with the Department of Trade and Industry, Department of Health, and MAFF – all of which touch upon matters of concern to the water industry. Parliamentary Committees that consider environmental matters and whose recommendations give a strong steer to the options and policies that UK negotiators advance, are also important, as is the Royal Commission on Environmental Pollution. The Water Services Association usually submits evidence to any inquiries they conduct of relevance to the water industry.

European channels of influence are also pursued, as follows:

- *The European Commission.* Because the Commission is bombarded with opinions, the Water Services Association channels its views through EUREAU and the European Waste Waters Group to give these views greater weight. However, it does not hesitate to go it alone if there is a particular British angle to put across. As well as reacting to current proposals, the Water Services Association tries to put forward positive suggestions, recognising that the Commission has the duty to initiate proposals for new Community policy. Efforts are also made to keep Directorates-General other than that for the Environment in touch with its views.

- *The Council of Ministers.* Agreement by Ministers is a late stage in the preparation of European law, but crucial for those major issues of principle which still remain unresolved. Where the interests of the Water Services Association would be adversely affected, a direct approach is made to the relevant UK Minister. More generally, contact is maintained with the Office of the UK Permanent Representative (UKRep). This is the part of the Foreign Office permanently based in Brussels to look after British interests in all aspects of European policy, and is an especially important channel in helping to establish which issues require debate by the Council of Ministers.

- *The European Parliament.* This has substantial powers under the 'new procedures' of the Single European Act and Maastricht Treaty and cannot be ignored. Any environmental opinion is prepared by a rapporteur and discussed in the Parliament's Environment Committee, so direct contact with UK and other members of this Committee is a worthwhile investment. In order to assist the level of parliamentary debate, it is crucial to provide MEPs with good and timely information and opinion.

- *The Economic and Social Committee.* The Committee's tripartite structure – with representatives of employers, employees and other interests – provides direct access for industrial interests who can nominate members and experts to serve on the Committee and its task forces. Its meetings are attended by Commission staff, allowing them to be cross-examined on the aims and content of draft legislation. Sometimes the opinions delivered by the Committee can be influential.
- *CEN and CENELEC.*[1] These are the two European standards-setting bodies. They are assuming greater importance as the Commission looks to them to define and refine technical standards, for example in relation to best practice on sewage sludge. Input for British industry is arranged through the British Standards Institution.

While pursuing these EU lobbying channels, the water sector does not neglect other international fora that have a formative influence in the development of European environmental policy. The Organisation for Economic Co-operation and Development (OECD), for example, can be the starting point for EU legislative initiatives or the reference point for methodologies used in the development of environmental policies. Here, the Business and Industry Advisory Committee, accessible via the CBI, is a major means of input, being one of just two official OECD consultative bodies. Likewise, the United Nations Environment Programme and the Economic Commission for Europe, whose activities the European Commission closely follows, are accessible to industry via the International Chamber of Commerce.

The Water Services Association employs consultants in Brussels to keep abreast of developments and also uses its European trade associations to forewarn it. Intervention at the European level, if it is to have any chance of success, requires sustained effort, involving considerable time and resources. Efforts actually to influence developments at this level therefore tend to be restricted to those which will quickly and obviously impact on the business activities of the water industry, such as drinking water quality, or sludge on agricultural land. More general or longer-term issues are left to the CBI and UNICE. Even on issues where the water industry may have a fairly strong sectoral interest, such as that of environmental liability, it may not be able to devote sufficient resources to attend to them.

The issue that has commanded greatest attention recently is the revision of the Drinking Water Directive. The extent to which technical and industrial expertise has been mobilised in this exercise and the openness of debate contrast sharply with the preparation of the original Directive (see earlier). Through the Water Services Association, the UK water sector has been particularly active in the process, seeking to improve the Directive's practicability and efficiency. It contributed in 1993 to a detailed analysis by EUREAU of the existing Directive. This provided a full report of how the Directive had worked in practice and what improvements to it were necessary. A European conference with the Commission and a series of national conferences, including one in London, involving local regulatory agencies and the Commission, took place to explain EUREAU's report and to press the case for the Directive to be revised. When the Commission submitted its proposal to the Council and the European Parliament in January 1994, it contained several features to improve its operation as well as taking on board the most recent scientific knowledge from the World Health Organisation. Since then, the UK water sector has maintained close contact with the Commission, and with the various research teams chosen by the Commission to examine specific parameters in the proposed Directive. It has also contributed to the work of the European Parliament and the Economic and Social Committee in preparing their Opinions on the proposal and has given both written and oral evidence to the inquiry carried out in the UK by the House of Lords Select Committee on European Legislation. For a Directive which will continue to determine a significant proportion of water suppliers' investment, all avenues to influence its operational flexibility have been explored.

There are areas where the water industry can indeed point to success, not least the revision of the Drinking Water Directive, which was in part attributable to the sustained campaign by EUREAU. Other successes include the recognition that the Urban Waste Water Treatment Directive would lead to greater generation of sewage sludge whose land-based outlets would need to be secured by, for example, avoiding its classification as 'hazardous'; and greater focus on the protection of groundwaters as an essential source of drinking water. So although the business community has brought about significant improvements such as these to certain environmental measures, it has done so through focused lobbying in relation to a limited number of issues, rather than across the board.

CONCLUSIONS

Environmental policy development in the European Union and the business response to it have evolved steadily and are now substantially different from when it all started in the early 1970s. The attitudes of all concerned have changed: the defensive, reluctant-to-compromise, abstract approaches at the start have been eroded as the Commission, government and business have all recognised that environmental protection and economic growth are mutually reinforcing goals. There is now a refreshing discussion of environmental objectives and how best to achieve them in a cost-effective way. Ever-tightening regulation is no longer taken for granted as the only means of securing environmental improvements (see Chapter 1).

European environmental policy has given a great stimulus to the environmental protection industry, of which the water industry is part, to develop new processes and systems. It has helped generate innovation in environmental modelling and impact assessment, in information exchange and in instrumentation and control technology. Even in process industry, where the direct costs of meeting new regulatory standards have been highest, gains in operational efficiency and lower raw material and waste disposal requirements have provided some degree of compensation.

European environmental policy has been good for British business and *vice versa*. The UK Government, encouraged by the business community, have succeeded in moving the Commission and other member states towards the needs for:

- consultation, for example to involve national regulators and those being regulated in the formulation of new regulations;
- better prior analysis of policy options and specific proposals, for example with a clearer scientific rationale or a risk assessment;
- integration, that is, to consider the environment in all its facets and in all Union policies;
- subsidiarity, that is, to set objectives and a framework for action at the European level but to allow action to occur more locally if necessary or desirable;
- better reporting mechanisms, so that compliance can be assessed more uniformly and consistently;
- better enforcement, so that member states or their industries do not benefit from the hidden subsidy of lax controls.

Obviously, in all these areas more still remains to be done: the picture is far from perfect, for example in applying in practice the 'polluter pays' principle. None the less, the trend is in the right direction.

Against that, other member states have moved the UK towards:

- understanding the 'Napoleonic' creed, whereby broad principles such as 'best practicable option' are built into the legal framework of regulation;
- accepting political compromises, even at some cost to business and against the balance of scientific opinion;
- debating means, other than controls, to achieve environmental goals, for example economic instruments or formal voluntary agreements;
- accepting futuristic 'targets', which may never be achieved but which are set so that member states have something to aim at;
- appreciating the benefits of working together as part of a major trading bloc.

Now that both the European Commission and the UK Government recognise the importance of involving business in environmental policy formation, the prospects of getting practicable and cost-effective solutions to priority problems look brighter than ever before. Of course, the real contribution of industry is in translating agreed policies and regulations into practice, via its investment and operating programmes. In the end, this is the way environmental improvement actually happens and is the true measure of the success of EU environmental policies.

NOTES

1 CEN is the Comité Européen de Normalisation (European Standards Organisation) and CENELEC is the Comité Européen de Normalisation Electrotechnique (European Standards Organisation for Electrical Goods).

Part IV

POLICY DEBATES: ISSUES, CONCEPTS AND INSTRUMENTS

Having covered in this book broad governmental strategies and the response of different types of institutions, Part IV examines the Europeanisation of specific policy fields, namely: waste management, nature conservation, land use and landscape, water quality and industrial pollution. For each of these policy fields, the respective chapter identifies the key characteristics of traditional British practice; the major European and legislative milestones; how Britain has influenced these; and how British practice has changed as a result. The aim of this part is to establish the consequences of Europeanisation for the substance of policy, the concepts and style of policy and the organisational relationships surrounding the policy process.

In Chapter 10, Andrew Jordan presents an overview of the impact on UK environmental administration. Over the past twenty years, he concludes, there have been profound changes. The EU's influence is easiest to discern in the precise setting of standards and the legal tools and instruments used. The effect on other aspects of policy has been indirect and contingent upon domestic developments, such as the growth in public awareness of the environment, the influence of the environmental lobby and the restructuring of government.

The individual policy chapters are written by experienced lobbyists and academics. They illustrate many of the generalisations made by Jordan, while revealing quite different patterns of UK–EU interaction. Martin Porter covers waste management in Chapter 11. This is a policy field which was just being established when the UK joined the EC. In consequence, there was scope for the UK both to exert influence upon, and to respond to, the emerging EC framework for regulating waste. Although there have been conflicts and tensions – not least over the priority the UK government gave to Single Market over environmental considerations on packaging issues – Porter characterises the overall experience as one of mutual 'policy learning' between the UK and other member states.

James Dixon covers nature conservation in Chapter 12 and here the starting point was quite different. The UK had had considerable experience in this field, possessed a well-established conservation movement and had played a leading part in international conservation. The EC presented a new arena in which to operate, not least for the voluntary organisations and not just to raise fresh issues but also to pursue the implementation of existing international conventions. Indeed, the development of EU legislation in this environmental field more than any other has been shaped by existing conventions. It is also the field in which Britain has played the most constructive role, due partly to the efforts of the conservation lobby but also to perceptions that action at the EC level would have little impact on established British procedures or commercial interests. Dixon does, however, identify internal tensions that have arisen within the conservation movement itself as the large voluntary organisations have established a strong European presence but the statutory agencies have struggled to find an effective role.

Land-use and landscape policy, covered by Fiona Reynolds in Chapter 13, is also a mature field but, in contrast, is one in which the UK and other governments have chosen to maintain national discretion for policy development and implementation. Except for agriculture, where the Common Agricultural Policy has been in place since 1957, land use and landscape remain the least Europeanised of the fields we are examining. However, Reynolds agues that pressures are building up from a number of directions – the need to reform the CAP, interest in protecting the European cultural heritage, the pressures on land use from economic integration – which will place land use and landscape at the heart of the debate about the future of environmental policy in Europe.

Water quality policy, analysed by Neil Ward in Chapter 14, was likewise a field where decentralised regulation was well and distinctively developed in the UK, but it has been transformed through the imposition of a European legal regime. Not surprisingly, this has proved to be one of the most turbulent and controversial features of EC–UK environmental relations. The impact has been inextricably bound up with the privatisation of water services, leading to fundamental organisational change as well as changes in the substance and procedures of policy, and the politicisation of the field.

Finally we turn to industrial pollution control in the chapter by Jim Skea and Adrian Smith (Chapter 15). Here, too, European pressures have contributed crucially to a revision of domestic policy, including reorganised structures and more formalised procedures. This is a field, however, in which the British Government's more recent efforts to play a leading role in EC environmental policy since the Maastricht Treaty have come to fruition in the new framework Directive on Integrated Pollution Prevention and Control. Moreover, throughout all these changes, the traditional British approach of site-level implementation has been preserved, leading Skea and Smith to pose the intriguing question: 'Is policy made through Brussels Directives or does it emerge through site-level implementation?'

10

THE IMPACT ON UK ENVIRONMENTAL ADMINISTRATION

Andrew Jordan

[T]he Communities have not penetrated dramatically into the national political scene but have . . . been confined predominantly within the executive (particularly within some departments) and within the sphere of national elites.

(Wallace 1971: 522).

Environmental policy in the UK is now inextricably bound up with EC policy . . . Much of the UK's environmental protection legislation is now developed in common with other EC member states.

(UK Government 1994b: 190)

This chapter seeks to identify some of the broader and longer-lasting effects of EU membership on the traditional principles, structure and style of British environmental administration. Along all three of these dimensions, the administration of environmental policy in the 1990s is manifestly different from what it was in the 1970s. The chapter tries to assess the relative influence of the EU in bringing about this transition, bearing in mind the various 'domestic' level factors which also modify policy. This in no way assumes a simple dichotomy between the 'national' and 'European' levels, for disentangling their respective influences upon the development even of a single Directive can be exceedingly difficult. For instance, European tools and concepts have been taken up by the UK Government and applied in areas not subject to EU policy. Nor does it imply that the flow of influence is necessarily from the EU downwards because British initiatives have on occasions helped fundamentally to shape EU policy. Rather, it provides a useful perspective for thinking about the impact of the EU on national political and administrative systems.

When Britain acceded to the Treaty of Rome in 1973 it accepted a body of law which formally commits all member states to certain common

173

policies, to defined procedures for reaching decisions on these policies and to enforcing their implementation in a consistent and coherent manner. What impact has this discipline had on the manner in which British officials enunciate and put into effect policies to protect and conserve the environment? The quotations above provide two helpful benchmarks although they say nothing about the intervening processes. The first, written by a leading academic analyst of European political affairs, describes the situation just prior to Britain's entry into the then EEC, which at the time had no environmental policy. While noting that the Community had eroded the traditional distinction between foreign and domestic policy, Helen Wallace concluded that 'European issues have ... tended to be regarded [in Britain] as one bundle of issues rather than as a new dimension pervading the political spectrum.' 'Changes at the national level', she continued, 'have not been extensive' (1971: 538). The second statement, which is drawn from the UK Government's strategy for sustainable development, issued in 1994, suggests that by then it had become increasingly difficult to conceive of or identify a free-standing entity called 'British environmental policy' distinct from EU policy; the two had become coterminous. Sbraiga (1996: 254) asks rhetorically how many other supranational organisations involve themselves in 'domestic' concerns like a country's sewerage system, its drinking water or the health of its trees. One is reminded of the prescient comment made by Lord Denning in 1974 in which he likened the Treaty of Rome to an 'incoming tide' which 'flows into the estuaries and up the rivers', and 'cannot be held back'. To carry this analogy further, if the effects of the EU are as profound as some commentators allege, we should be able to detect them not only in the rivers and streams of policy, but also in the creeks and rivulets that make up the British administrative system.

This is a view to which Nigel Haigh firmly subscribes, although he makes the important point that the impact of Europe varies even within a single policy sector such as the environment:

> not many areas of [UK environmental] policy have now been left entirely untouched by the [EU] even if the depth of involvement remains uneven. Some fields, such as the control of hazardous chemical substances, have largely been defined by [EU] policy. Others, such as pollution of air and water, while profoundly affected by [EU] concepts, retain distinctively national characteristics. In contrast, town and country planning ... has so far been much less influenced by the [EU].
>
> (Haigh 1994b: v)

Other chapters in this volume cover some of these areas in much greater detail. This chapter, however, concentrates upon the meta-policy level. It does so by disaggregating the notion of 'administration' into its constituent parts. For convenience, these could be loosely defined as operating *principles*, organisational and legal *structures* and the broader policy *style*. In reality, these three not only overlap but are to a great extent interdependent. The operating *principles* of policy are the assumptions which inform policy and determine not only the overriding goals of policy and the instruments used to attain them, but also the very nature of the problems they are meant to be addressing. *Structures* are the bureaucratic and legal arrangements used to deliver policy. Policy *style* is a broad term used to describe the characteristic operating procedures adopted by organisations and the agents which work in them. Some analysts claim that all countries have a characteristic 'style' of processing issues and problems (Richardson 1982). Europeanisation could be viewed as a force promoting the convergence of national systems, their principles, structures and styles. Do we see this occurring in the environmental field or does British administration retain an independent character?

CONTEMPORARY UK ENVIRONMENTAL PROTECTION: THE EMERGENT CHARACTERISTICS

Whatever the cause, it cannot be denied that the administration of environmental policy in Britain has undergone a subtle but profound revolution in recent years. Several *principles* of the emerging system stand out: an increasingly formal and uniform system of administrative control based upon fixed standards and timetables for compliance; a wider adoption of source-based controls as prescribed by the Continental notion of precautionary action; a greater concern, now enshrined in legislation, for the health of the environment and the intrinsic rights of non-human entities,[1] rather than just human health; and a greater attempt to make the polluter pay a larger portion of the social costs of pollution, starting with the administrative cost of permitting. In terms of *structure*, the trend is simultaneously towards a shift in the locus of control from 'front line' (mostly local) regulators to officials at higher levels, who in turn try to adopt a more integrated view of problems; a more complex system of relationships with many new players drawn into environmental regulation; and a greater willingness to experiment with non-regulatory forms of intervention such as environmental taxes. The predominant trends in the *style* of administration include: greater readiness to enunciate the underlying principles and objectives of control; a more open and independent style of regulation, now increasingly but by no means entirely conducted at arm's length from operators and service delivery agencies; and greater and more uniform monitoring, and a wider dissemination of information about the state of environmental quality.

However, to varying degrees, all political and bureaucratic systems possess powerful gyroscopic tendencies that resist even committed attempts to

Table 10.1 Key UK domestic environmental legislation since 1973

1973	Water Act
1974	Control of Pollution Act Health and Safety at Work Act
1976	Endangered Species Act
1980	Local Government Planning and Land Act
1981	Wildlife and Countryside Act
1982	Derelict Land Act
1985	Food and Environment Protection Act
1989	Water Act
1990	Environmental Protection Act
1991	Planning and Compensation Act
1995	Environment Act

overhaul and re-steer them. Thus, British environmental policy is still more secretive than that of many other countries; administrative discretion continues to prevail over judicial interpretation; there is still a strong attachment to informal voluntary agreements and non-quantified standards; the status of the precautionary principle remains equivocal both politically and legally (Hughes 1995); authorities remain reluctant to set clear and legally binding targets other than those specifically required by EU or international legislation; and authorities are still a long way from making polluters pay the full social costs of pollution, while environmental taxes are conspicuous by their absence in most fields.

Of course, in explaining these developments, overarching importance should not be ascribed to European influences when domestic and wider international factors would in any case have altered British structures and practices. It would be hard, for example, to explain solely in EU terms the provenance of the newly integrated Environment Agency, the publication in 1990 of the most comprehensive statement on environmental matters by a British Government, the loosening of ties between scientists and national officials, the opening up of environmental and health standard-setting procedures, and profound shifts in British policy on issues such as acid rain. Of the 'internal' catalysts, the most salient are *inter alia*: the enormous growth in environmental awareness among the public, business and the media; the growing size and sophistication of environmental pressure groups; the intellectual input of the standing Royal Commission on Environmental Pollution; and the restless urge of successive Conservative administrations to overhaul and reform British government. Meanwhile, the steady internationalisation of the environmental agenda during the 1980s and 1990s, encompassing issues such as acid rain and North Sea pollution, and major conferences such as the 1992 Earth Summit in Rio, have precipitated important shifts in national policy as Britain has been obliged to comply with international conventions and burden-sharing agreements with its neighbours and trading partners. All these factors would have influenced domestic environmental policy regardless of the UK's membership of the EU.

The difficulty lies in teasing out the 'European' from the 'domestic'; or, to be more precise, understanding how European obligations have been processed and refined by domestic actors and a British political system that possesses its own internal dynamic.[2] What we can never know is the counterfactual, namely what would have happened had the UK not joined the then EEC. None the less, a number of commentators, including senior civil servants (Osborn 1992), Ministers (Waldegrave 1985), distinguished scientists (Ashby and Anderson, 1981), environmental lawyers (Macrory 1991), Commission officials (Krämer 1995) and policy analysts (Haigh 1986; 1994a), have tried to assess the broader effects of EU membership on the process and content of British environmental policy. Abstracting from their accounts, the EU has allegedly:

- *forced the UK to defend its overall approach to environmental protection and justify the underlying principles of action, many of which were arcane, implicit or merely rhetorical.* High-profile disputes between the UK and the Commission might have been resolved in a way which allowed the UK to preserve important elements of independence, but they nonetheless forced Ministers to provide a clearer explication of policies and informing principles that were either taken for granted or the product of political and economic expediency. The EU has thus challenged the UK's 'administrative sovereignty' (Krämer 1995: 154) and geographical isolation, obliging it to consider and on occasions adopt Continental principles and practice, and bring forward changes in policy faster than it might otherwise have done;
- *helped to centralise into the hands of national and supranational officials responsibilities for setting standards, determining priorities and making investment decisions that were once exercised at the local or regional level.* Nowadays, policies relating to a whole range of issues, some of which are almost exclusively local in their cause and effect, are routinely processed at the international level by central government Ministers, who, of course, are ultimately responsible for ensuring that EU laws are properly enforced and implemented. Membership of the EU has thus forced national government to become much more closely involved with local affairs;
- *kept up the pressure for higher environmental standards and accelerating the pace of remedial work.* The extension of the EU's involvement has created profoundly important constraints on the power of member states. It has

177

Plate 10.1 Tall chimneys dispersing emissions from Britain's power stations have been blamed for contributing to acid rain in northern Europe. The Scandinavian countries were the first to call for international action to curb such emissions in the 1970s, but the Community only began to respond in 1982 after concern erupted over the death of forests in Germany. The Commission put forward what became the Large Combustion Plants Directive (88/609), which broke new ground by including total national emission limits subject to phased reductions. Britain, as the largest emitter of SO_2 and heavily dependent on coal, was initially strongly opposed to the measure, which took five years to negotiate.

Source: Greenpeace/Gazidis

forced on to the domestic political agenda issues such as drinking water that were previously neglected or actively downgraded. On the implementation side, the EU has constrained the Government's freedom to pursue independent policies, disavow promises or defer expensive clean-up operations. In short, the EU has succeeded in drawing a 'line' in the sand from which successive UK Governments, harried by environmental groups, have found it difficult to retreat. Water quality policy is a good case in point (see Chapter 14). That 'line' is of course negotiable as the dispute between the Commission and the UK over bathing water designations clearly shows, but not to the same extent as with domestically inspired policy;

- *obliged British officials to operate in transnational networks.* When a recent British Foreign Secretary remarked that he met with his opposite numbers in other EU states more frequently and for much longer periods of time than his Cabinet colleagues, he revealed an important shift in the centre of gravity of British government. After a period of hesitation, British civil servants in the Department of the Environment (DoE) have learnt to operate in European policy networks and to adapt to joint decision-making procedures that increasingly require building alliances with other states (see Chapter 2). It is commonly remarked that Europeanisation has reduced the power of British officials to revise British policy. Less well appreciated is the DoE's greater ability to influence the environmental policies of other member states (Haigh 1995: 14). Meanwhile, officials at the local authority level and within statutory agencies have also fostered links with their opposite numbers in other member states (see Chapter 5);

- *created a more explicit and transparent framework of environmental protection that has reinforced the trend towards greater openness in regulation.* European legislation tends to embody fixed numerical standards and deadlines to ensure comparability of effort and simplify the process of monitoring and enforcement. The UK, on the other hand, has traditionally relied upon voluntary agreements with polluters, general guidelines, informal standards and implementation systems that can be tailored to suit political and financial exigencies (see Chapters 14 and 15). The Union's attempts to harmonise standards and abatement methods across Europe have helped to erode the UK's attachment to a case-by-case or contextual approach to control, whereby standards are attuned to local conditions and circumstances;

- *encouraged greater environmental monitoring and helped to increase public understanding of environmental quality.* Water and air quality are now more regularly and uniformly monitored (often at considerable public and private cost) than they ever were in the past. The information derived from monitoring, which traditionally was retained within tightly drawn policy communities of national officials and technical elites, is submitted to the

Commission and released into the public domain, where it has formed the basis of successful pressure group campaigns and, on occasions, cases considered by the European Court;

- *brought the UK into contact with new tools and principles of environmental protection and influenced the manner in which it applies existing tools.* Community directives have *inter alia* supported the precautionary principle[3] and the wider application of technology-based controls at source. The UK has been obliged to give administrative devices such as water quality objectives legal backing and to approach industrial permitting on a more consistent, sector-by-sector basis (see Chapter 15). Completely new tools such as air quality standards have been introduced as a direct result of EU law. Meanwhile, the Commission is struggling to inaugurate a strict liability regime which will oblige polluters to pay a greater portion of the costs of remediation consonant with the polluter pays principle. Traditionally, the UK has tended to rely upon common law remedies of public and private nuisance.

The rest of the chapter considers these claims and seeks to delineate the implications for the principles, the structures and the style of British environmental administration.

THE PRINCIPLES OF UK ENVIRONMENTAL ADMINISTRATION

'Discretion and practicability might be described as the hallmarks of British environmental law and policy, with a degree of satisfied isolationism and administrative complacency running closely behind' (Macrory 1987). Britain is credited with having the oldest system of environmental protection of any industrialised society. The Alkali Inspectorate was originally established in 1863 to regulate acidic emissions from the chemical industry, while the Rivers Pollution Prevention Act of 1876 made it an offence to pollute river water. The former introduced the concept of 'best practicable means' (BPM) which remained the cornerstone of air pollution control until it was replaced in 1990 by the European equivalent BATNEEC (Best Available Technology or Techniques Not Entailing Excessive Costs) (see Chapter 15). As new problems emerged and became important, new laws were enacted and new agencies put in place. The approach was predominantly reactive rather than anticipatory, tactical rather than strategic, pragmatic rather than ambitious, and case-by-case rather than uniform. These traits, which are still replicated across many areas of British policy making, led to a somewhat jumbled mixture of organisations, principles and legal tools. Certainly until the publication of *This Common Inheritance*, the 1990 White Paper on the environment, the received view was that:

government structures and law relating to environmental protection
have been an accretion of common law, statutes, agencies and policies. There is no overall environmental policy other than the sum of
these individual elements.

(Lowe and Flynn 1989: 256)

But running through these responses was a particular way of thinking. 'All
authorities', stated a 1978 review of pollution control in Great Britain (DoE,
1978), 'are expected to operate on the philosophy that standards should be
"reasonably practicable".' This somewhat arcane concept has been broadly
defined as:

> the use and maintenance of equipment and the operation and
> supervision of processes in such a way as to ensure that any dis-
> charges . . . are controlled as far as is practicable, having regard
> among other things to local conditions and circumstance, to the
> current state of scientific knowledge and medical knowledge of the
> potential harm or nuisance involved, and to the financial implica-
> tions. This pragmatic approach permits the establishment of indi-
> vidual standards for polluting emissions . . . which can be made
> continually more stringent in the light of technical advance and of
> changing environmental needs, but allows greater flexibility than
> statutory standards.

(*ibid.*: 3)

The costs of compliance and the actual threat of harm were thus seen as
important factors in setting emission controls. Gradualism was the dominant mode of action: controls were strengthened according to economic circumstances, scientific knowledge and the goodwill of producers, not
imposed uniformly. The 'contextual' approach was staunchly and proudly
defended by the scientific and political establishment – the standing Royal
Commission on Environmental Pollution and the influential House of
Lords Select Committee on the European Communities. Lord Ashby, a
highly distinguished member of both, conceded that the British approach
might appear to be 'the product of expedience, not principle: a policy to be
described as a non-policy' (Ashby and Anderson 1981: 153). Even so, he
insisted that it was 'the product of nearly two centuries of evolution in
which the impracticable ideas have been eliminated, Utopian aspirations
have been discarded, and the policies which have survived have been proved
to work'.

In an illuminating study of acid rain, Martin Hajer (1995) shows how the
British response was strongly conditioned by the beliefs of influential members of the policy elite such as Ashby, who were proud of Britain's record
of dealing with environmental problems and convinced of its superiority

181

over alternative (read Continental) approaches. Such practices as setting uniform standards for emissions or target groups were regarded as less efficient, over-legalistic and unjustified on scientific grounds. This insight helps to explain some of the problems that the UK has experienced in adapting to joint environmental policy making and Continental axioms of control: conflict was inevitable. With no implementing agents of its own and limited resources, formal regulation is the only credible policy tool available to the Community to attain its environmental objectives. Standards and timetables have to be explicit to encourage consistency of application and to permit monitoring and external review. Haigh (1992: section 3.9), however, argued that the difference in doctrine was never that sharp since there was an element of uniform standard-setting in the British approach (for example, for certain emissions to air). But more often than not, British standards generally took the form of informal objectives with quaint titles such as 'presumptive limits' or 'normal standards'. They were administrative devices, not fixed standards enshrined in legislation. Thus drinking water had to be 'wholesome', emissions to air 'harmless and inoffensive' (although only where practicable), and so on. In the late 1970s, water quality objectives were set but they were purely administrative devices with no binding force. Interpretation was left to local officials who took pride in their technical knowledge and understanding and their ability to interpret policy in the light of local circumstances.

While the flexible-contextual approach might have enjoyed the support of those charged with administering it, others, including the Commission and fellow member states, were understandably suspicious. Although it might have served an island state like the UK admirably well, it was plainly too unsystematic in terms of burden sharing and far too reliant upon local interpretation and enforcement to serve as a model for the Community as a whole. In 1977, Michel Carpentier, then head of the Commission's Environment and Consumer Protection Service (later to become DGXI), captured this sense of mistrust in a nutshell:

> [W]e certainly recognise the long experience of Britain in environmental matters . . . But . . . incredible as it may seem the rest of the Community do not necessarily agree on the absolute paramountcy of axioms which are close to the heart of British experts and officials. For example the 'Continentals' tend to believe more in standards defined on the basis of best technical means and applied through mandatory instruments. They mistrust systems based on goodwill and voluntary compliance . . . They have serious doubts about the absorptive capacity of the environment . . . They also tend to believe that where the scientific and medical evidence is inconclusive, we should err on the side of caution.
>
> (quoted in Levitt 1980: 93–4)

Within Britain, misgivings were also expressed by those outside the elite. One former Assistant Secretary at the DoE admitted that the public 'can understandably become alarmed' when seemingly lax discharge conditions in certain localities are offset by tighter ones elsewhere (Renshaw 1980: 237). For critics like Chris Rose (1990: 4) of Greenpeace, flexibility and informality 'proved a recipe for fudge and smudge, a quagmire of intellectual fuzziness and a licence for administrative laxity, which allowed the progressive deterioration of the British environment'.

Underlying the British approach is a specific view of what constitutes an environmental 'problem'. For the British, controls should be set according to the extent to which a particular substance or practice actually damages the environment, but especially human health. Given that some level of waste emissions is inevitable, so the thinking goes, it seems prudent to make full use of the environment's innate capacity to absorb and render waste harmless (Waldegrave 1985: 108). On this line of reasoning, it is assumed that pollutants (substances that cause damage) can be differentiated from contaminants (those wastes that could be safely assimilated), although clearly this is not always the case. Following this approach, broad quality objectives are set for different areas according to the use to which they are put (one might not, for example, insist upon the same level of protection for a river used for drinking water abstraction as one used for fishing) and individual discharge consents set appropriately. In other words, the 'tendency is to focus less on eliminating pollutants at source than restricting to an acceptable level their effects after they have entered . . . the environment' (Royal Commission on Environmental Pollution 1984: 46). The leitmotifs are balance, flexibility and pragmatism. According to Ashby the British 'believe it is more effective to select the areas most at risk and concentrate one's efforts upon them' (Hansard, 13 October 1975, vol. 364, col. 725). Attempts to impose standards uniformly will inevitably leave some areas over-protected and some under-protected; hence his famous remark about 'let[ting] the fish and the shellfish decide . . . [what] is favourable for them' (Hansard, 28 June 1977, vol. 384, col. 1070). The locus of monitoring should therefore be the environment rather than the point of discharge.

Uniform standards existed in Britain for certain types of products such as noise from cars, but they were the exception rather than the rule. The broader aim is not to attempt to eliminate waste emissions entirely or reduce environmental damage to zero, but to seek an *optimal* balance on the basis of an assessment of costs and benefits (Ashby and Anderson 1981: 136). This requires a certain flexibility in standard-setting; rigorous and unambiguous scientific proof of actual damage is therefore regarded as the *sine qua non* of effective control. In practice, however, both costs and benefits were more likely to be defined in relation to the more narrow circumstances of the polluter than those of society as a whole. The operational manifestations of the contextual approach, which gained the unfortunate epithet 'dilute and disperse', are the tall stacks and the long sea outfall pipes, whose primary purpose is to externalise waste. Why did policy elites in the UK favour such an approach? Part of the answer lies in the country's geographical isolation (the availability of strong rivers, fresh westerlies and active tides to disperse emissions) and high population density. The tradition for civil servants to rely upon scientists for specialist advice was also a strong contributory factor.

Conflicts between British and Continental thinking emerged as soon as the EU began setting formal standards for particular substances, habitats and levels of environmental quality in an attempt to harmonise controls and remove barriers to free trade. The most famous and certainly the most protracted centred on the emission of dangerous ('red list') substances to water. After a long and acrimonious struggle,[4] driven as much by Britain's determination to protect the interests of its domestic industry as to defend the intellectual coherence of its approach or improve environmental quality *per se*, a compromise was eventually found and incorporated into European law (see Chapter 14).

The process of adaptation has at times been painful and protracted. Underlying many of the early disputes was an insistence that the UK should shoulder rather than externalise its own burden of environmental damage – something which the long pipes and tall stacks were manifestly not designed to achieve. On many issues, the UK only accepted stricter environmental standards when confronted with irrefutable scientific evidence of harm. For this it gained a not entirely deserved reputation for being an awkward European partner and an environmental laggard. Although the UK continues to make a strong contribution to international diplomacy, it still preaches the virtue of taking action at the national level, typified by its support for the principle of subsidiarity (which it invariably interprets as supporting action at the national rather than EU or sub-national levels) and for the review and refinement (read emasculation) of especially troublesome pieces of legislation such as the Bathing and Drinking Water Directives.

To what extent, then, has the philosophy of environmental policy altered in recent years? Certainly the Government has tried to enunciate a more *explicit* statement of principles, first in the 1990 White Paper (UK Government 1990: 10)[5] and more recently in its strategy for sustainable development (UK Government, 1994b: 32–4). Hall (1993) suggests that policy change should be assessed at three distinct levels. First-order changes concern the setting of policy instruments. Here there clearly has been much EU-inspired change to UK practice. Later chapters document several instances where existing standards have been tightened as a result of European pressure.

Next there are second-order changes which are about the nature of the instruments themselves. Here there has been some continuity but also change, with much greater adoption of source-based controls and fixed emission and environmental quality standards. The secular trend in the UK is away from generalised standards or implied objectives towards more coherent and explicit standards set according to predetermined criteria and formal processes of assessment. So, for example, there is a special category of SSSIs related to the Birds and Habitats Directives where the Government has less leeway to allow development (see Chapter 12). The Environmental Impact Assessment Directive requires a more formal assessment of the potential impacts of development on flora and fauna as part of the traditional consideration of 'other material considerations' in the planning process (see Chapter 13). Likewise drinking water must still be 'wholesome' but detailed regulations – almost all deriving from European legislation – provide authorities with technical criteria upon which to base their judgements (see Chapter 14).

Most fundamental of all, however, are third-order changes to the overarching goals or principles that guide policy. Hall believes that change at this level occurs infrequently, as a result of a basic shift in thinking, such as that from Keynesianism to monetarism. The equivalent in the environmental field would be a transition from the 'traditional-pragmatist' approach (Hajer 1995) to one based on the notion of precaution. Now a fundamental goal of Community environmental policy, the principle of precaution calls for action to be taken to protect the environment in advance of scientific proof, for prevention (that is, source- or process-based control) over remedial action and for the traditional focus on human health to be supplemented with a concern for non-human entities. Although these are all alien to the British tradition of practicability, flexibility and 'sound science' (O'Riordan and Jordan 1995), the UK Government has expressed its support for the precautionary principle, but with the crucial qualification that it should only be invoked when there are 'significant risks of damage to the environment' and 'the balance of likely costs and benefits justifies it' (UK Government 1990: 11). Thus the principle is seen to apply 'particularly when there are *good grounds* for judging either that action taken promptly at *comparatively low cost* may avoid more costly damage later, or that irreversible effects may follow if action is delayed' (emphasis added) (*ibid.*: 11). This looks suspiciously like 'practicability' in a new guise.

The UK is still not an instinctive or unqualified supporter of precautionary measures and it is this difference in outlook which separates it from the greener member states and the Commission.[6] Of course, the British regime for air pollution control always contained an element of precaution, but unambiguous instances of the principle's application in recent years are hard to pinpoint, not least because the principle itself is unclear. Haigh (1994b: 242) believes that the UK's climate change strategy is a good example of precautionary action, although the House of Lords Select Committee on Sustainable Development is more circumspect in its judgement (1995: 14). The responses to stratospheric ozone depletion and North Sea pollution are also commonly cited as examples of its application, but these too are debatable claims. Meanwhile, UK officials continue to preach the merits of the contextual approach in European fora: the UK, for example, strongly opposed the application of rigid emission limits in areas such as integrated pollution prevention and control (IPPC) (see Chapter 15) and battled to maintain lower standards of sewage treatment for more 'robust' coastal waters during the negotiation of the Urban Waste Water Treatment Directive (see Chapter 2). The principle of best practicable environmental option (BPEO),[7] which is one of the central pillars of the UK's integrated pollution control (IPC) regime, is strongly informed by the contextual approach.

THE STRUCTURES OF UK ENVIRONMENTAL POLICY

The tradition in Britain has been for legislation to provide a broad framework within which local officials could treat problems on a case-by-case basis. Even then, legislation was only used as a backstop, 'when other methods [had] failed' (Waldegrave 1985: 106). Rather than setting firm standards, Parliament passed laws incorporating open-ended concepts such as 'best practicable means' (BPM) or 'as low as reasonably achievable' (ALARA) which had to be interpreted in local contexts. This fitted with the contextual philosophy of relating general objectives to local conditions and circumstances. Until the practice was rebuked by the European Court, the UK transcribed European Directives in much the same way, using shadowy mechanisms such as administrative circulars and technical notes that could be changed when it was expedient but which lacked legal force and were therefore not open to legal challenge. Since rights of private prosecution were limited (under the statutory law, action was reserved to public bodies), enforcement was in the hands of local officials who saw recourse to the courts as a sign of failure rather than success. A strong tradition developed in pollution control, nature conservation and land-use planning of avoidance of legal action if at all possible. In general there has been a strong attachment to voluntary agreements with industry rather than formal regulation. Self-regulation has also been encouraged, most particularly with farming interests (Cox, Lowe and Winter 1990).

Because there was so little legal intervention, key concepts such as 'practicability' were not the subject of open and authoritative legal interpretation. Best practicable means (BPM), for instance, was not defined in statute until 1956, and has never been interpreted by a higher court, and is now unlikely ever to be so following the repeal of the Alkali Act in 1996, which closed a chapter of history stretching back more than 130 years. In the absence of guiding first principles, legal systems developed in an *ad hoc* and 'uncoordinated manner' (Macrory 1991: 9). This was in keeping with the British tradition of common law with its interplay of precedent and gradual adjustment to change, and the lack of a rigid constitution. EU law, on the other hand, follows the pattern of Roman law which tends to grow 'top down' by the imposition of decrees embodying precise rules and requirements. Macrory opines that European laws have provided an 'underlying dynamic for change' that is carrying the UK towards a more 'formal' legal system, embodying explicit goals, statutory objectives and clearer criteria for making administrative decisions. Together with public registers of information and more open rights of private prosecution, this is creating a 'potent mixture unfamiliar to British practice' (*ibid.*: 19).

But has the EU been the only source of change? Not really. International conventions would in any case have introduced fixed goals and targets, but they would not have reached as deeply into the domestic scene or arrived with such regularity. Larger businesses, too, have realised that tougher and more independent regulation plays well with customers, employees, shareholders and potential investors because it indicates that they are employing best practice techniques. Indeed, many are beginning to divulge information voluntarily about their processes beyond that required by legislation, although smaller companies still have a long way to go. The more general growth in environmental awareness and concern has also been a factor, creating a public demand for a more explicit demonstration that environmental standards are being set and complied with. The 1990 White Paper responded by introducing a system of target setting and review at the policy level, although it has been roundly criticised by environmental groups for lacking ambition, and doubts persist about its full implementation (*Independent on Sunday*, 17 March 1996). None the less, targets or indicators of biodiversity conservation, waste generation and disposal, and sustainable development have either been recently published or are promised. The Environment Act 1995 obliges the DoE to prepare a National Air Quality Strategy with standards, objectives and measures to be taken by different authorities (for example, the designation of Air Quality Management Areas). However, environmental groups, supported by the House of Lords Select Committee on Sustainable Development and the Government's Advisory Panel on Sustainable Development, have a much more challenging process in mind, something akin to the Department of Health's 'Health of the Nation' initiative which sets long-term, numerical goals and assesses progress against them.

If legal systems are becoming more transparent and formal, what implications are there for traditional practices and procedures? Clearly, much depends upon the use that different actors make of laws, targets and standards. The widely predicted rise in the volume of environmental litigation is yet to appear. Court proceedings are still expensive and judicial review procedures long, expensive and uncertain as to their outcome. The rules regarding *locus standi* are normally sufficiently tight to dissuade many potential litigants from pursuing action for the public interest. The courts, meanwhile, seem reluctant to involve themselves in technical issues. There has been much talk within the legal profession of a specialised environmental court but nothing has yet come of it. Likewise the Commission, supported by the House of Lords Select Committee on the European Communities, has long promised to introduce rules allowing citizens greater access to national courts to circumvent the often protracted process of seeking justice in the European Court, but again, no action has yet been forthcoming. That British environmental groups have often found it more productive to complain to the Commission is a good indication of the constraints they face in the UK.

In terms of the *organisation* of environmental protection, successive governments acted on the principle that problems should be dealt with at the local or regional level wherever possible. Complex issues that could not effectively be dealt with locally (for example, acidic emissions, radioactive waste and vehicle emissions) were managed centrally or left to specialised authorities such as the Alkali Inspectorate. Otherwise, a decentralised structure was consistent with the contextual approach, which clearly could not have been applied by bureaucrats working at desks in London, Cardiff or Edinburgh. What emerged from over a century and a half of gradual experimentation and adjustment was a fragmented system of locally based controls and standards. Water authorities, for example, determined their own investment priorities within spending limits set by the Treasury, while individual standards were determined by local officials who used the broad reference standards as a basis for negotiation (Hawkins 1984).

However, in the 1980s a succession of Conservative Governments pushed through a massive programme of institutional change: the civil service was slashed, nationalised industries privatised and various publicly run functions contracted out or shifted to 'Next Steps' agencies. Measures of output and performance, explicit standards and incentive structures, external audit and regulation of service delivery, and closeness to the consumer were the principles which informed these reforms. From areas like education to environmental protection, local authority control was systematically reduced, while many of its service delivery functions, particularly in the waste disposal field, were contracted out (Jenkins 1995; Chapter 11).

These changes have touched all policy areas, environmental protection included, although the practical outcomes have been somewhat contradictory. The national conservation body, the Nature Conservancy Council, for example, was broken up in the late 1980s into country agencies (see Chapter 12), whereas environmental protection functions have gradually been centralised, first in the National Rivers Authority and Her Majesty's Inspectorate of Pollution, and now within an integrated Environment Agency for England and Wales (a separate Scottish Environmental Protection Agency has been set up). With an annual budget of £550 million, a staff complement of nearly 10,000 and offices across the country, the Environment Agency will be the largest and most powerful of its kind in Europe. Its Chief Executive promises a more preventive approach rather than prosecution, and to communicate directly with the public via education campaigns. Two-thirds of its budget will be paid for out of permitting fees and administrative charges. It will take over the waste regulatory function from local authorities and various other responsibilities (among them contaminated land, and genetically modified organisms) from central government. Meanwhile, the appearance of regulatory watchdogs in the water, electricity and gas sectors, of which none has a specific remit to consider environmental issues, has ushered in a much more open and formal style of price and investment regulation. In the past five years, both OFWAT and OFGAS have openly challenged the basis of environmental policy decisions on waste water treatment and climate change respectively, and seem to be developing their own notion of the public interest in fighting for the interests of the consumer (Maloney and Richardson 1995). In the water sector the head of OFWAT, Ian Byatt, has pushed determinedly for a more open and publicly accountable system of financial regulation, and has called upon central government to lay bare the financial calculations underpinning standards, many of which were brokered by Ministers at the European level (see Chapter 9). This arrangement stands in sharp contrast to the traditionally closed and locally based forms of regulation described above.

Crucially, most of these developments are domestically inspired. Claims, therefore, that EU Directives have directly centralised powers in Britain (Haigh, 1986) or forced the DoE to take on a greater policy 'steering' role should properly be interpreted in the light of these other trends. Some change would have occurred irrespective of the EU as regulation became steadily more international in scope. It is, however, true that EU Directives are now negotiated, adopted and transposed at the national level by central government Ministers and officials in association with their equivalents from other member states. Environmental and conservation agencies and local government play a secondary role in implementing what has been agreed; and their role in influencing and interpreting policy has been diminished (see Chapter 5), although, as Morphet shows, local government interests at least have sought to counteract this by establishing their own links with the EU (see Chapter 8).

THE STYLE OF UK ENVIRONMENTAL ADMINISTRATION

Writers such as Richardson (1982) claim that the dominant *style* of British government is informal, reactive, gradualist and accommodative. With Jordan (Jordan and Richardson 1983) he describes it as a system of 'bureaucratic accommodation' among close-knit and tightly drawn policy communities of interested specialists and specialist interests which cluster around government Departments. Great emphasis is placed on consultation and negotiation, rather than imposition and confrontation. This in itself militates against anticipatory action and radical policy change, a tendency we have noted above. According to Richardson, all policy problems are processed in this way. Weale (1992: 81) likewise claims that in the UK:

> [t]here is a desire to avoid programmatic statements or expositions of general principles governing particular areas of policy. The preference is for the particular over the general, the concrete over the abstract and the commonsensical over the principled.

Explanations for this include the absence of a rigid constitution, the existence of a large, diverse and very well established pressure group system and the 'generalist' tradition within the civil service (which itself generates a need for advisers to inform and legitimise policy decisions). Environmental regulation, which tends to have a particularly strong technical component, displays many of these traits. There has generally been a long-standing willingness on the part of both the public and politicians to leave such matters to professional inspectors and officers. In operational terms, regulation has tended to be based upon courteous negotiation between actors who see themselves as members of a club of like-minded individuals rather than adversaries. As Vogel (1986: 77) revealed in his comparison of UK and US environmental regulation, there is a long-standing reluctance to set standards with which compliance cannot be guaranteed. It is bound up with the whole notion of 'practicability' and to a large extent it lives on in the attitude of Ministers who negotiate at the European level. There, the UK is commonly seen as a recalcitrant negotiator but a fairly dutiful implementor of legislation. It is, moreover, commonly remarked that British pragmatism helps to make EU legislation more workable, and British officials like to think they inject a dose of common sense into what are often highly rhetorical or abstract debates.

Instead of signing up to policies which it either cannot or will not implement, the UK has a reputation for carefully scrutinising proposals to ensure that they are workable. As one Minister, David Trippier, remarked during a spat with the Commission over alleged breaches of EU law in Britain:

> In one Member State, everything is permitted unless it's forbidden. In another Member State everything is forbidden unless it's permitted. And in some member states everything is permitted – especially if it's forbidden.
>
> (ENDS 1991: 2)

However, the manner in which the UK has chosen to couch its criticisms and articulate its ideas has sometimes been misinterpreted as a deliberate attempt to be stubborn and 'awkward' (George 1990). It is often said that the problems the UK has experienced in adapting to Continental practices boil down to an inappropriate style of operation in Europe rather than deriving from fundamental disagreements with the substantive issues under discussion (Wallace 1995).

Whatever the case, the adversarial practices favoured in the USA and parts of Continental Europe – using standards to 'force' technology, judicial scrutiny and litigation – have been largely absent in the UK. The keystone of the whole approach was, and to a large degree remains, *trust*. To nurture and protect it, the rights of individuals to take private prosecutions were statutorily curtailed and public access to information restricted in a way that protected the confidentiality of emitters. Studies have shown that this approach inevitably led to the regulators adopting too readily the standards and approaches of the regulated (Richardson *et al*. 1983), although in his comparative study Vogel (1986: 23, 26) concluded that the style of regulation had less of an impact on outcomes than geographical conditions, or economic and technical factors; and that the British style was ultimately no less effective than a more open and adversarial approach.

Even so, the style of environmental regulation in the UK is undergoing a metamorphosis. Again there are elements of continuity and change. We have noted the trend to more explicit standards, the diminution of local discretion and the centralisation of control. The turnover of staff in the new Environment Agency alone is likely to lead to the replacement of the old cadre of pollution control specialists inculcated in the traditions of 'practicability' and negotiated consent. These developments have all perturbed the quasi-secretive world of pollution control as has the advent of public registers and mechanisms of judicial review. McAuslan (1991) foresees the emergence of a 'more open, discursive, but not necessarily confrontationist approach more along the lines of the planning system'.

The EU's effect on these developments seems to have been contingent rather than direct. The explicitness of EU legislation and the Commission's preference for more uniform targets, report and review mechanisms have undoubtedly helped to open up the British system of regulation to greater external scrutiny and reduced the discretion once enjoyed by local officials. For instance, the degree of sewage treatment, once a matter for local decision, is largely prescribed by the Urban Waste Water Treatment Directive. At the same time, though, both industry and the environmental movement have increasingly pressed for more explicit and consistent standards. In the final event, however, their implementation rests largely on negotiation between regulators and the regulated. Indeed, after a brief flirtation with a more arm's-length approach in the early 1990s, Her Majesty's Inspectoraate of Pollution found that it was quicker and more realistic to employ a policy of quiet, but arguably more structured, negotiation than prescription in implementing IPC (ENDS 1996c; Chapter 15).

In terms of reducing the culture of administrative secrecy in the UK, the EU's role has also been somewhat limited. Since it first raised the issue in its second (1972) report, the Royal Commission on Environmental Pollution, backed by a number of pressure groups, has waged a long battle for greater openness. A Directive designed to improve public access to environmental information held by public authorities was only adopted by the EU in 1990 and implemented in the UK in 1992, although many individual Directives require states to monitor and systematically report upon their implementation. However, the data generated are fragmented, limited in scope and often difficult to interpret. In the meantime, openness and accountability have emerged as issues of party political competition within the UK. John Major's personal mission to open up the machinery of government to public scrutiny via the Citizen's Charter and Open Government White Paper (Cabinet Office 1993) initiatives furnished much factual information but failed to clarify the precise lines of accountability within government. In opposition the Labour Party responded with proposals for an American-style access to information Act. This may prove to be a far more powerful source of change to the culture and style of British government than anything currently under consideration in Brussels, which of course has its own problems with secrecy and unaccountability!

CONCLUSIONS

It is difficult to deny that the traditionally 'British' system of environmental protection has undergone a quiet but profound transformation over the past twenty years, prompted by a whole range of different factors. Isolating a specifically European component from the general ferment is, however, exceedingly difficult and probably impossible. The EU itself has changed greatly in the more than two decades since the UK acceded to the Treaty of Rome, and many of the significant initiatives emanating from the development of the EU's environmental policy would seem to have been pressing against an opening door in the UK. That said, the EU's influence is easiest to discern in the precise setting of standards and the legal tools and instruments used. On the other hand, its effect on the principles, style and organisational structure of environmental legislation has been indirect and largely contingent upon other domestically inspired developments. This assessment is consistent with the dominant mode of EU action. Directives, and to an extent Regulations, set out the objectives to be achieved but leave member states to determine the precise means of achieving them.

It would, of course, be mistaken to see the EU solely as an external, 'top down' influence on the UK, although this might have been more true in the 1970s and 1980s when the doctrinal disputes between the UK and the Community were at their height. We have, for example, barely touched upon areas where Europeanisation has been less profound, such as land-use planning, or where the flow of influence and new thinking has been in the opposing direction (for example, the integration of environmental factors into other policy areas, IPPC and nature conservation). In addition to the discretion routinely built into EU legislation, the UK remains free to alter, at times radically, the organisational structures that deliver environmental protection, which in turn has implications for the style of regulation. The intriguing question raised by Haigh – referred to at the beginning of this chapter – is why convergence has proceeded faster in some sub-sectors of British environmental policy than others. The process of adjustment was always likely to be more painful and protracted in areas such as water and air pollution, with well established institutional structures and traditions in place, than areas such as waste disposal, in which UK policy has largely co-evolved with European policy (see Chapter 11). Member states continue to assert their sovereignty in the field of land-use planning, where policy making is still subject to unanimous voting.

Perhaps the best way to think about Europeanisation, then, is as a process of mutual learning and adjustment between two co-evolving political systems. However, the direction and extent of learning has altered over time. In the 1970s, a combination of naïveté among policy elites about the operation and importance of the Community and a more general feeling of satisfied isolationism, meant that the UK contributed little by way of initiative to EC policy and in fact often found itself on the defensive. The fact also that most policy was, until 1987, adopted via unanimous voting allowed the UK to adopt such a reactive stance. Since then, however, the wider application of majority voting, a series of initiatives by more openly pro-European Secretaries of State (Chris Patten and John Gummer), and a growing realisation of the need to engage in constructive dialogue with other actors at *all* stages in the policy process (particularly agenda setting and policy shaping), have brought British thinking and practice closer into line. Indeed, there is now clear evidence of a convergence in policy orientations as subsequent chapters show. After a frenetic twenty-year period of policy development, EU environmental policy has arguably entered a period of consolidation, brought about by the troubled ratification of the Maastricht Treaty and the deep economic recession of the early 1990s. Currently the leitmotifs of policy are subsidiarity, simplification and integration across policy areas, and there are even strong voices calling for deregulation. Ironically, this has created a context which is more conducive to traditionally British notions of practicability and flexibility. The future legislative programme has, for instance, been slashed while the Commission has come under strong pressure from various quarters to ensure that new proposals are clear in terms of the objectives to be achieved and justified by a proper consideration of costs and benefits. It is even considering the merits of voluntary agreements with industry. Not surprisingly, the UK remains a strong supporter of full and more consistent implementation of existing standards – a recurrently awkward bone of contention among member states. At the same time, there are even indications – for example, in the Directive on IPPC and the

Commission's Communication on water resource management (CEC 1996c) – that the once polarised debate about the respective merits of the 'contextual' and uniform standards approaches may finally be reaching a compromise which recognises the benefits of both.

ACKNOWLEDGEMENTS

Tim O'Riordan, Neil Ward, Rob van der Laan, Heather Voisey, Philip Lowe and Stephen Ward all provided helpful comments on earlier drafts of this chapter. Responsibility for remaining errors and omissions rests entirely with the author.

NOTES

1 The Environmental Protection Act (1990) requires operators to use the BATNEEC to prevent, minimise or *render harmless* polluting emissions.
2 See Nigel Haigh's (1984: 9–25) useful discussion of the links between national and Community policy.
3 The precautionary principle demands that action be taken to prevent environmental damage even if there is uncertainty regarding its cause and possible extent, on the grounds that it is better to be roughly right in due time, bearing in mind the potentially costly consequences of being very wrong, than to be precisely right too late. Precaution emerged in the former West Germany in the 1970s, but has since become an accepted principle of European Union and international law (O'Riordan and Jordan 1995).
4 Most commentators tend to concentrate upon the struggle over the 'black list' (of more polluting) substances. What is often neglected is that *all* member states, the UK included, committed themselves to setting emission standards for the 'grey list' (less polluting) substances according to water quality objectives. This was the first time that the environmental quality objective approach became mandatory in the UK, although the National Water Council had begun to consider the use of river water quality objectives in the mid-1970s in order to systematise the setting of discharge consents. In this sense, the EU accelerated a process of change that was already occurring in the UK.
5 These are contained in the first chapter: to use the best evidence available; to apply the precautionary principle where appropriate; to ensure the public is informed; to work for international co-operation; and to use the best instruments of policy. The opening chapter was apparently written by Chris Patten himself.
6 The dispute about dangerous substances was largely about how much precaution should be exercised: the greener member states wanted strong precautionary measures (the best available technology) applied at the source of emissions, whereas the UK argued for a contextual approach with standards based on the characteristics of the receiving water. The latter implicitly assumes that thresholds of damage can be ascertained, which is not always the case.
7 The BPEO is 'that option that provides the most benefit or least damage to the environment as a whole, at acceptable cost, in the long term as well as in the short term' (RCEP 1988: 5).

11

WASTE MANAGEMENT

Martin Porter

The waste management sector has experienced a considerable change since the early 1970s when the UK acceded to the then EEC. Standards have been tightened; policy has become more comprehensive, integrated, proactive and prescriptive; and administration has become more centralised, less informal, more open and encompassing. However, it would be a mistake to attribute most of these changes to the influence of the EU alone. Other factors have been at play during this period, ranging from wider international pressures to changing environmental circumstances and domestic political factors. Nevertheless, there has undoubtedly been symbiosis between EU and UK developments in which influence and impacts have been reciprocal and variable over time.

On the one hand, there are several instances where the UK has influenced EU waste management policy, including setting the EU agenda and providing a model on which to base policy. There are also many instances where it has negotiated important changes to policy proposals to bring the eventual legislation more in line with prevailing UK policy. It has also influenced EU policy evolution on occasions by advancing arguments which have re-oriented some waste issues from primarily environmental matters into Single Market ones and *vice versa*. Opinion varies as to whether these changes have been 'negative' or have injected some much needed 'realism' into EU waste management policy, but they certainly have had important impacts.

On the other hand, there are also many ways in which EU policy has had an impact on the substance and concepts of UK waste management policy and even on its administration and procedures. Changes in domestic law and practice have had to be made to incorporate EU legislation. However, often the effect has been to speed up what would probably have occurred in any case or to oblige better implementation than would otherwise have been the case. Although some standards are undoubtedly somewhat higher and administration more centralised than would have been so, the most significant consequence of EU membership for the UK appears to reside in the fact that central government must respond to the innovative and pressing agendas of other member states. Through a process of 'policy learning' it thereby assimilates experience and thinking from other member states. One implication is that an innovative and proactive approach will often be the one which reaps most benefits for the UK within the EU, even if it presents some initial difficulties at the domestic level for a system that is inherently conservative.

UK WASTE MANAGEMENT POLICIES AND ADMINISTRATION FROM THE 1970s TO THE PRESENT

The UK's efforts to deal with waste management in a comprehensive way date back to the early 1970s. Growing domestic concern with the environmental effects of waste had prompted the government to establish working groups in 1964 and 1967 to examine the issues of toxic waste and refuse disposal. The resultant reports, which appeared in 1970 and 1971 respectively, at a time of intense international concern over the growing evidence of global environmental degradation, laid the basis for Part I of the 1974 Control of Pollution Act which restructured UK administration and policy on environmental issues more generally (Weale 1992).

The Control of Pollution Act introduced an innovative waste management system which required waste disposal authorities (WDAs) to draw up plans for the management of all non-toxic, domestic, commercial and industrial wastes, giving details of the quantities accruing, disposal methods, the costs involved and so on. WDAs (county councils in England, district councils in Wales and Scotland) were required to consult various interested parties, including water authorities and other local government tiers, and to publicise draft plans to enable public representation on them to be made. The final plans had to be notified to the Department of the Environment (DoE). The legislation also introduced a comprehensive licensing system, administered by WDAs, for facilities disposing of these non-toxic wastes. Under this scheme, each WDA was obliged to maintain a public register of disposal licences. The 1978 Refuse Disposal (Amenity) Act placed a further obligation on local authorities to provide suitable sites for local residents to dispose of large items of waste free of charge.

Different provisions applied to toxic wastes which were already regulated under the 1972 Deposit of Poisonous Waste Act which had been introduced earlier following a high-profile scare about the dumping of such waste (Haigh, 1984). Anyone removing or disposing of toxic waste was required to notify the WDA and the water authority at least three days before doing so. This Act also generated considerable data about industrial waste because the 'notifiable' wastes were defined 'negatively', in effect including all waste not specifically designated as non-toxic. However, when the Act was repealed and replaced by the 1980 Control of Pollution (Special Waste) Regulations,

introduced within the framework of the Control of Pollution Act, the definition of such 'special waste' was loosened by using an approach which combined an 'inclusive list' with other criteria. Local authorities gained additional
responsibilities under these changes.

Waste management in the UK had long been a responsibility of local
government (Haigh 1984). Before the 1974 Act, the powers and duties of
local authorities to control waste as part of their public health responsibilities under the 1936 Public Health Act ensured that it was their representatives (environmental health officers) who were charged with detecting and
removing 'noxious matter' and for prosecuting those responsible for its
deposit. It was also local authorities who, under the 1947 Town and Country
Planning Act, could give planning permission to new developments, such as
waste disposal plants. Under the Control of Pollution Act, the key role of
local authorities continued. Although central government did have some
reserve powers – for example, to adjudicate in disputes between a WDA and
an applicant for a licence – it was essentially concerned with developing an
overall policy and issuing guidance.

The characteristics of the relationships between those involved in waste
management during this period appear to be typical also of other areas of
UK environmental policy (see Chapter 1), variously characterised as 'a
reactive and consensual policy style' (Richardson and Watts 1985: 17) and as
striving to be 'technically effective and informal' (O'Riordan and Weale
1989). In such a 'low politics' approach, there was in waste management
issues the close and closed consultation with affected interests (site operators, industrial and commercial concerns with wastes to dispose of, water
authorities, and so on) typical of the period.

Consensus and pragmatism were also apparent in the way in which policy,
once agreed, was put into practice and enforced (Heidenheimer et al. 1990).
Local authorities and their inspectors refrained from adopting a legalistic
and confrontational stance towards those industries producing wastes,
unlike their opposite numbers in, say, Germany (Richardson and Watts
1985). Rather, they pursued what might be called negotiated compliance;
infringements, if detected, were resolved most frequently without taking
matters to court. In any case, the scope for legal confrontation was greatly
diminished by the way in which the WDAs combined regulatory and operational functions, being responsible for both checking up on all locally generated waste and disposing of much of it, too. This combination of
responsibilities did, however, allow some difficulties to be swept under the
carpet; moreover, the lack of central governmental involvement in
enforcement meant that policy implementation was far from uniform. One
result of such decentralised regulation and technical discretion was that
there was a remarkable amount of variation between local authorities in the
specification of landfill sites, even for hazardous wastes (Madel and Wynne
1987).

To a large extent this was a deliberate result of the UK's preference for policy implementation responsive both to local environments and to the cost implications. However, waste management policies also tended to be piecemeal, emphasising single-medium discharge rather than a multi-media approach. The waste issue was mainly perceived as a question of how to handle and dispose of waste safely, that is from a human health point of view, and cheaply. Environmental concerns were secondary and were conceived largely in terms of avoiding public nuisance and damage to amenity. The issues of landfill availability, problems of water contamination from waste sites, air pollution from incineration, international trade in wastes and resource consumption issues were not on the governmental agenda. Indeed, even an environmental group such as FoE, which initiated its campaign against non-returnable bottles in 1971 and emphasised the importance of recycling throughout this period (FoE 1990), would probably not have fully anticipated the changes in public and official perception of the problems of waste management which have since occurred.

Notwithstanding the fact that the UK has had to implement several important pieces of EU waste legislation since 1975 (see below), the key policy and administrative changes to UK waste management have come more recently, in the wake of the Environmental Protection Act 1990. The genesis of this legislation can be traced to a critical report by the Royal Commission on Environmental Pollution (1985). A significant administrative development in response was the re-organisation in April 1987 of three environmental inspectorates (the Industrial Air Pollution Inspectorate, the Hazardous Waste Inspectorate and the Water Quality Inspectorate) into Her Majesty's Inspectorate of Pollution (HMIP). Even at that stage observers commented on the fact that this was not only an effort to unify the somewhat disparate environmental controls which then existed but was part of a discernible trend in national environmental policy away from traditional policy content and style, towards more openness and information, a more comprehensive approach to policy (part of which was multi-media waste management) and more structured accountability (O'Riordan and Weale 1989).

In addition to the creation of HMIP, there were several changes relevant to waste management, also suggested by the Royal Commission, which the government accepted in principle in a series of consultation and discussion papers (DoE 1986; DoE/Welsh Office 1988; DoE/Welsh Office 1989). These included: a 'duty of care' obligation on the producers of waste; the introduction of a registration scheme for waste transporters; additional powers for WDAs concerning site licensing, enforcement and aftercare of disposal sites; and statutory backing for the national Hazardous Waste Authority (Haigh 1995b). Also included were an enhanced role for central government in the setting and monitoring of standards and an important change in the role of local authorities which separated the functions of regulation and operation. Some EU influence is apparent in these changes but less so than in other aspects of environmental policy such as air pollution (Haigh 1989; 1995b).

The Environmental Protection Act distinguishes three different types of waste authority: Waste Regulation Authorities (WRAs), Waste Disposal Authorities (WDAs) and Waste Collection Authorities (WCAs). The key functions of WRAs are the drawing up of waste disposal plans, and the supervision and inspection of regulation. They do not carry out disposal themselves but award contracts either to private companies or to ones set up at arm's length by local authorities. WDAs may also direct WCAs on waste collections. Lastly, WCAs must respond to these directions and draw up and implement plans for waste recycling. These complex administrative arrangements mean that local authorities may in some cases still have regulatory, disposal and collection responsibilities, but the provisions of the Act mean that the functions must be kept separate to avoid the 'game-keeper turned poacher' problem. Following the Environment Act (1995), this set-up changed again with the establishment of the Environment Agency. The new agency combined the functions of HMIP, the NRA and WRAs and therefore took over some of the roles of local authorities as regards waste disposal. More significantly perhaps in the longer term, the Environment Agency also has a role in developing and implementing a national waste strategy, something which has only recently become part of governmental policy and which represents a further move away from piecemeal and reactive approaches towards comprehensive and anticipatory ones and towards greater centralisation.

The first intimations of this national waste strategy came in 1994 when the government published its response to the 1992 Rio Earth Summit in the form of a National Strategy for Sustainable Development (UK Government, 1994b) which included a chapter on waste. This partly answered the criticism that a number of disparate and incremental initiatives of the previous four years needed a coherent overall plan. Specific policies and targets were not outlined but the objective of shifting waste management practices further up the waste hierarchy (see Figure 11.1) was expressed. Greater clarity came in December 1995 with the publication of a White Paper on the subject (DoE/ Welsh Office 1995). This outlined the general waste hierarchy (reduction, re-use, recovery and disposal), with the adoption of best practicable environmental option (BPEO) to determine specific applications. Within this hierarchy, there is no preference for recycling and composting over incineration and no targets are set for minimisation or re-use. There are, however, targets for recycling 25 per cent of household waste by 2000, recovering 40 per cent of municipal waste by 2005 and reducing the proportion of controlled waste consigned to landfill from 70 per cent to 60 per cent by 2005.

The national waste strategy is to be a statutory strategy and is due to be prepared by the DoE under advice from the Environment Agency. In order

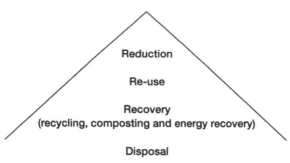

Figure 11.1 The waste hierarchy

to do this, the Environment Agency must complete a national survey of what waste is being generated and life-cycle analyses to identify BPEOs for particular waste streams. With a greater emphasis on market-based instruments, such as the landfill tax (a weight-based tax on landfill waste which came into effect in 1996), to achieve targets, and greater promotion of the issues and solutions involved, the consequences are expected to amount to a 'quiet revolution' in UK waste management practices over the next few years (ENDS 1995b).

From a procedural point of view, the creation of the Environment Agency is undoubtedly a centralising measure, even though its Chief Executive has been at pains to emphasise the 'hands-off' and devolved way in which it hopes to operate (Gallagher 1996). This is the inevitable consequence of greater central governmental involvement in the setting of policy as well as in its implementation. The poor track record of local authorities in developing and implementing waste disposal plans was certainly a contributory factor in this: Haigh (1989: 136) suggests that even by 1987, 65 of the 198 plans which should have been prepared under the 1974 Control of Pollution Act were still incomplete. But in addition to other domestic and bureaucratic factors – such as those identified by O'Riordan and Weale (1989) in the creation earlier of HMIP – there was also certainly some pressure to move in this direction in order to ensure compliance with EU environmental legislation.

In other respects too, the way in which policy is now made and administered is considerably different from the situation of the early 1970s and may represent a qualitative shift in the accepted 'policy style'. The setting of targets, though still quite limited, is nevertheless something of a sea-change because it allows more precise measurement of success or failure from a public point of view. The impact of this is quite considerable. It enables public interest groups to hold the government to account more easily than was the case, making both the goals of waste management and their achievement (or otherwise) matters of public scrutiny and debate rather than the outcome of private consultations in restricted policy communities. It also means that a somewhat more legalistic, less pragmatic, approach to policy implementation is likely. Already, the way in which HMIP went about enforcement had been more activist than its predecessors (*Independent*, 3 October 1991; *Financial Times*, 18 September 1993). HMIP's enforcement operations, though, had been hampered by a lack of resources and staff shortages: initially, only 5 of the 200 employees were waste disposal inspectors and they had to rely on persuasion of the 79 WDAs rather than direct enforcement (*Financial Times*, 2 December 1988; 2 September 1989). Enforcement powers were enhanced under the 1990 Environmental Protection Act and, despite its relatively streamlined administration, the Environment Agency is likely to continue this trend towards increased activism in enforcement.

The way in which the Environment Agency applies the new 'producer responsibility' principle will be indicative in this respect: whether it restricts itself to advice and persuasion, or is prepared also to use the legal means available to it to see that the principle is enforced. The 'producer responsibility' principle is also revealing about the evolution towards greater integration of waste management issues into mainstream environmental and economic practices. The way in which waste is now being dealt with 'in the round' rather than as an 'end-of-pipe' problem does not only affect regulators. It is also having considerable impacts on the businesses responsible for producing waste. The wider involvement of businesses at all stages of the production, distribution and supply chain, as well as of consumers, has greatly widened the number of interested parties involved in consideration of waste management policies. The UK's attempts to develop a coherent response to the problem of packaging waste have involved consultations with consumer and environmental groups, raw materials producers, packaging converters, packaging users, distribution companies, wholesalers and retailers, waste disposal companies and local authorities. Although use of the best possible scientific advice remains at the forefront of government approaches, this greater public awareness and participation will add to the range of factors considered by policy makers in drawing up and implementing policy. Rather than relying mainly on scientific and expert consensus, the attitudes and opinions of the general public may become more important, especially given the importance of their participation in any policy if it is to succeed in achieving its environmental objectives. For example, the readiness of consumers to purchase plastic packaging, the willingness of citizens to engage in recycling and the degree of public acceptance of incineration all have an impact on the ability of government to achieve its targets.

Many of these changes in policy have undoubtedly come about because of a greater urgency to find solutions to some environmental problems faced by the UK, such as the diminishing availability of landfill sites, and accumulating evidence of other environmental problems, such as leaching from landfills and the impact of toxins released by incinerators. Reasons for policy change may also be found in technological advances in the field of waste management, domestic political priorities and programmes and wider international developments (such as the Rio Earth Summit), all of which have influenced public perceptions of the issues and the practicability of solutions to them. None the less, it is important also to recognise that some of the key influences in this progression, while not always acknowledged publicly, have come from EU initiatives and from obligations and pressures on the UK inherent in its membership of the EU. Is it possible to tease out this particular dimension?

KEY EU WASTE MANAGEMENT POLICIES AND LEGISLATION

Although the EU had no legislation concerning waste before the mid-1970s, since then it has developed a considerable range of policies which now deal with most aspects of waste management (see Table 11.1). During the 1970s, it introduced the Framework Waste Directive (1975) and so-called daughter Directives on Waste Oils (1975), PCBs (1976) and Toxic Waste (1978). Waste management was also a theme in the Second Environmental Action Programme. The 1980s saw the adoption of legislation concerning the Transfrontier Shipment of Toxic Waste (1984; 1985; 1986; and 1987), Sewage Sludge (1986) and Beverage Containers (1985) and a Commission Communication on the EU's waste management strategy (1989), something supported by the Third and Fourth Environmental Action Programmes.

The 1990s have seen substantial revisions to, or replacement of, all of the pre-existing legislation, together with significant additional initiatives. The Framework Waste Directive was amended in 1991 and those dealing with the shipment of waste were consolidated in 1993. A 1991 Directive on Hazardous Waste in effect replaced the 1978 directive, although this was delayed by the late arrival of additional legislation concerning a list of hazardous waste (1994). Legislation has also been adopted concerning certain 'priority waste streams' (packaging – which replaces the 1985 Beverage Cans Directive – and batteries/accumulators), and on Urban Waste Water Treatment (1991) (see Chapter 14), Integrated Pollution Prevention and Control (see Chapter 15), and PCBs (polychlorinated biphenyls) and PCTs (polychlorinated terphenyls). It is expected soon on landfill and incineration standards, and is possible on liability for damage caused by waste. Additional waste streams which have been the subject of consideration, but not yet legislation, include used tyres, end-of-life vehicles and electrical, demolition and health care wastes. Uncertainty also still surrounds the Commission's

Table 11.1 Key EU waste management legislation and policy

CEC (1989), A Community Strategy for Waste Management. SEC(89)934
Council of the CE(1990), Resolution on Waste Policy (C122 18.5.90)
CEC (1996) Review of the Community's strategy for waste management
  (COM(96)399)

Waste – Framework Directive: 75/442/EEC
Waste – Framework Directive amended: 91/156
Waste – List: Commission Decision 94/3

(Directive on hazardous waste (78/319) repealed by Directive 91/689)
Directive on hazardous waste (91/689)
Directive on hazardous waste amendment (94/31)

Regulation on the supervision and control of shipments of waste within, into and out
  of the European Community (259/93) (replaced directives 84/631, 85/469, 86/279
  and 87/112)

(Directive on the disposal of polychlorinated biphenyls and polychlorinated
  terphenyls (76/403) repealed by Directive 96/59)
Directive on the disposal of PCBs and PCTs (96/59)

Directive on the disposal of waste oils (75/439)
Directive on the disposal of waste oils – amendment (87/101)

Directive on batteries and accumulators containing certain dangerous substances
  (91/157)
Directive adapting this (marking) (93/86)

(Directive on containers of liquids for human consumption (85/339) repealed by
  Directive 94/62)
Directive on packaging and packaging waste (94/62)

Directive on the protection of the environment and in particular of the soil, when
  sewage sludge is used in agriculture (86/278)
Directive concerning urban waste water treatment (91/271)

Directives on the prevention and reduction of air pollution from municipal waste
  incineration plants (89/369 and 89/429)
Directive on the incineration of hazardous waste (94/67)

Directive on integrated pollution prevention and control (96/61)

Proposed Directive on civil liability for damage caused by waste (COM(91)219 Final)

---

views on how to progress a proposal on environmental liability which, like
the more specific one dealing with waste, has lain dormant after arousing
much discussion in the early 1990s.

A Council of Ministers Resolution in 1990 welcomed and supported the
Commission's Waste Management Strategy. Developed within the Fifth
Environmental Action Programme, it has as its objective the 'rational and
sustainable use of resources' and sets out the following goals: stabilisation
of the amount of waste generated at the 1985 level of 300 kg per capita

each year, recycling or re-using 50 per cent of paper, glass and plastics; and the ending of exports outside the EU for final disposal by the year 2000. At the centre of the EU's strategy has been the waste management hierarchy outlined in the 1989 communication. This emphasises the avoidance of waste as a first priority, followed by the maximum amount of re-use and recycling, and then the safe disposal of anything else in the following order: combustion as fuel; incineration; and finally landfill. The precise interpretation and realisation of this have been subject to considerable controversy, evident especially in the debate surrounding the packaging and packaging waste directive (Porter 1995). Other policy principles, now outlined in the EU's waste strategy, include the so-called proximity principle which suggests that waste should be disposed of as close as possible to its source and the self-sufficiency principle which suggests that each member state establish a network of disposal installations capable of dealing with all of the waste generated in that state.

UK INFLUENCE ON EU WASTE MANAGEMENT POLICY AND LEGISLATION

The UK influence on EU waste management policy has been variable. In the early years it was quite apparent and the UK was keen to be seen as a role model. However, during the 1980s in particular, the agenda was increasingly determined by considerations of other member states with different concerns. Proposals that emerged sat less easily with the prevailing policy and ambitions in the UK. During this period, UK influence on and strategy towards EU waste management policy was reactive and conservative, though not inconsequential. In the 1990s, although the pace of change has still been in many ways dictated by Germany, the Netherlands and Denmark, as well as the new Nordic member states and Austria, the UK has been more constructive and innovative itself, helping to set the agenda and providing useful examples of new approaches. This changed orientation – by no means restricted to waste management but more generally applicable to environmental policy – has resulted in a greater influence on policy development and legislation which now more often sits relatively comfortably with UK concerns and policy style.

There are three principal ways in which one can analyse the influence of the UK on EU waste management policy: first, whether a national initiative provided the impetus and model for the EU's policy thinking; second, whether, during negotiations, changes were made to the policy in response to UK representations; and finally, whether following agreement, challenges were made to the form and content of the legislation and, either through negotiated amendments or through a ruling from the European Court of Justice, these led to changes in policy. Although analytically distinct and useful for that reason, in practice these are not wholly separate categories, as the case of the Framework Waste Directive and its amendments will illustrate.

The Framework Waste Directive, agreed in 1975, would seem to be a clear example of EU policy influenced by a piece of UK legislation. Although there were also French and German initiatives occurring at the same time, the UK Minister responsible claimed that the 1974 Control of Pollution Act provided the model for the Directive, an assertion supported by Haigh (1984). The lack of changes to the proposal during negotiations indicate the lack of difficulty which the UK felt it had with the Directive. Largely because of problems with its implementation, an amendment was proposed to tighten the definitions and to strengthen certain provisions. Again the influence of the UK is evident, in two ways, in the subsequent developments. First (admittedly with other member states), it objected to the Commission's attempt to base the amended Directive on Article 100a. This dispute (case 155/91) was eventually resolved by a ruling from the European Court in March 1993 which stated that Article 130s was the appropriate legal base in that instance because the primary objective of the Directive was to protect the environment while the trade and competition implications of it were ancillary. The UK was also responsible for raising the issue of self-sufficiency – an important plank of the amended Directive – at a Council meeting in September 1989, and for an amended definition of waste (Haigh 1995b).

The 1978 Directive on Hazardous Waste probably stemmed from French and Belgian initiatives, even though the UK claimed that once again the Control of Pollution Act was used by the Commission as the model (Haigh, 1989). During the negotiations on this Directive, the influence of the UK, alongside other governments, is apparent in amendments which led to the transport of toxic waste being omitted from the scope of the Directive, a qualification to the definition of toxic waste and restrictions on the power of the technical progress committee. However, it was unsuccessful in regard to another amendment which would have made it easier to recycle such waste. Negotiations surrounding the amendments to this Directive which were agreed in 1991 (and further amended in 1994) were relatively unimportant from the point of view of UK influence, although disputes over the waste list are revealing: the UK (together with Italy) was opposed to the list drawn up, on the grounds that it was too broad; and the exemptions introduced to accommodate this view were so limited that they both voted against it but without being able to block the proposal.

The UK influence on the first of the various Directives on the shipment of waste may be seen in the publicity given to certain shipments between the UK and the Netherlands, although these were only a few among several incidents that prompted EU action (Pestellini 1992). Although the provisions of the Directive were largely welcomed by the UK, it did seek amendments concerning waste destined for recycling and liability, and on both counts it was successful. The 1993 Regulation was prompted mainly by the

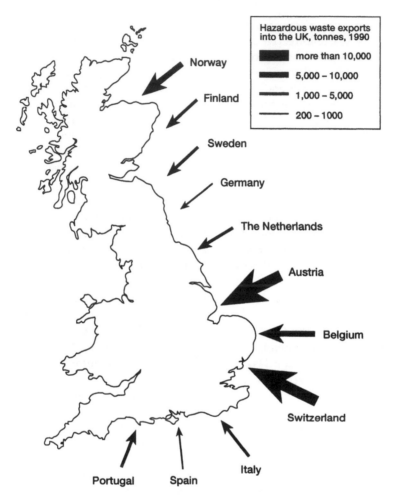

Figure 11.2 The origins and relative volumes of hazardous waste exports into Great
        Britain
Source: Stanners and Bourdeau (1995), p. 753

need to give effect to the provisions of the Basel Convention and OECD
decisions on transfrontier movements of waste, but it also provided the
opportunity to replace the original measure which had suffered from patchy
implementation in the run-up to the official inauguration of the Single Market. The incentive to regulate effectively was given further impetus by a 1992
ruling from the European Court in case C-2/90 over the Wallonia Waste
Ban. The UK was also keen to see its initiative on self-sufficiency adopted
and, supported by France, argued strongly in favour of this. Although the
final Regulation was a compromise designed to accommodate the worries of
smaller member states, the UK was content with it (Haigh 1995b).

UK influence in the so-called Beverage Cans Directive of 1985 was not so much in the agenda-setting (which came mainly from a piece of Danish legislation making standardised returnable containers for beer and soft drinks compulsory) as in the negotiations and subsequent changes which eventually resulted in the 1994 Packaging and Packaging Waste Directive. In both cases, the concern of the UK was to ensure that the EU legislation preserved the Single Market and that the environmental obligations were not too onerous for its industry and administration. Thus, in several instances it is possible to see the UK influence on the progress of the draft beverage cans proposal from its earliest form in 1980 to the final Directive: legally binding targets were dropped, for example, and measures were introduced which meant that any member state intending to introduce legislation in the field would have to report it to the Commission for vetting over its compatibility with the Single Market. The outcome was a Directive which was criticised by industrialists and environmentalists: too weak to be effective from an environmental point of view but also so vague in its provisions that the Single Market was far from safeguarded from divergent national policies. However, the UK and much of its industry was pleased that some of the over-ambitious suggestions of early drafts had been removed. These sorts of modifications have been variously described as 'negative' and typical of Britain's 'Dirty Man of Europe' image (Rose 1990), or as introducing a 'healthy dose of realism' (UK Government Ministers at the time) into EU policy making. Whatever one's perspective on the changes, the UK influence is clear.

The UK also played a key role in the way in which the Danish legislation was subjected to criticism on Single Market grounds. Not only were UK business interest groups, such as the Industry Council on Packaging and the Environment (Incpen), particularly vociferous in their complaints against it, but the government backed the Commission in its decision to take the Danes to the European Court. Although the ruling largely vindicated the Danish position and thus went against the Commission, and by extension the UK's view, it paved the way for more legislation at the EU level. This was because one part of the ruling made it clear that the Danish legislation was acceptable only in the absence of comprehensive EU legislation. Against a background of poor implementation of the Beverage Cans Directive and the unilateral initiatives being taken not just by the Dutch but by the Germans also, the EU initiative may thus be seen as much as a Single Market measure as an environmental one, and indeed, the legal basis for the Packaging and Packaging Waste Directive tends to confirm this view (Porter 1995).

The UK was, therefore, not especially active in influencing the environmental issue at the EU level, but in seeking an EU response to the adverse economic effects of national environmental initiatives. Although it had the backing of several other member states in this regard, the UK was particularly prominent in arguing this case during the discussions over the proposal, in which endeavour (and with some irony) it was helped by the chaotic impact of the German packaging ordinance on packaging waste exports within and beyond the EU. Consequently, the final Directive actually reflects UK interests to a large degree: the targets for recovery and recycling were lower than originally proposed and were broadly in line with the objectives which the UK Government had by then set itself; there were no specific targets for waste minimisation or re-use; the waste management hierarchy was not strict but indicative, in line with UK views; and the scope for member states to exceed the provisions of the Directive was limited by clauses ensuring that they had sufficient domestic capacity to deal with the recyclate they had collected. While the Dutch Minister complained that 'for the Single Market it is a good proposal, but not for the environment. It has nothing to do with the environment', the comments of the UK Minister Tim Yeo on this Directive were predictably positive: 'Commonsense has prevailed . . . the Directive has been transformed from one driven by green zealots.'

By contrast to the emphasis on Single Market issues which can be seen in the packaging issue, the agricultural sewage sludge issue had been identified as a clear environmental concern even before the EU had an environmental policy and before UK accession in 1973. Although the UK had a strong interest in the development of the Directive because of the large amount of such sludge generated and used on agricultural land, it was opposed to many of the conditions in draft proposals which it considered too rigid and insufficiently cost-effective. However, four years of negotiations led to a Directive which met a great many of the UK's objections (Haigh 1995b).

A similar pattern of events may be identified in the 1991 Urban Waste Water Treatment Directive (UWWTD), although in this case there was a major change in policy from the UK Government during the negotiations which made its agreement to key parts of the Directive possible (see Chapter 2). The Directive did not hit the EU agenda due to any British pressure but originated in environmental and health concerns over the effects of discharges of untreated sewage into fresh and coastal water, notably the Adriatic (Von Weizsäcker 1995) and in the waters of Germany, the Netherlands and Denmark (Haigh 1995b). The early drafts were very demanding and would have involved enormous investment in a very short period in order to ensure the treatment of virtually all municipal sewage. While some changes were introduced which met UK concerns (but which also responded to the interests of Southern member states) one would have expected the UK to oppose a key point in the Directive: the ending of sewage dumping at sea by 1998. However, on the eve of the Third North Sea Conference, the UK announced that it would phase out this practice. The reason for this about-turn was a combination of diplomatic pressures from other North Sea states and the Commission, together with increasingly negative domestic public opinion over the issue of dumping (see Chapter 2).

The costs involved would be sizeable but the UK government was more relaxed about this than it might otherwise have been because water privatisation meant that the cost implications would be borne by private companies and their customers. UK influence is therefore apparent in the greater flexibility achieved in some aspects of the final Directive, but must be set against the fact that the agenda had been driven by other states and that, probably for EU and other international negotiating reasons, it accepted legislation which in other circumstances it might not have done (Porter 1996).

One should also consider the UK influence on other recent developments. The IPPC Directive, for example, was inspired by the British approach which was developed during the 1980s and which came into effect more properly in the 1990 Environmental Protection Act (see Chapter 15). By contrast, the Directives on the incineration of hazardous waste and on landfill have their origins more in German and Dutch concerns. That said, the UK has been able to negotiate some changes which bring them more in line with its policy objectives. Overall, therefore, one can identify periods in early and more recent policy where the UK has been influential in setting the EU agenda and providing useful models for EU policy. More often, however, influence has been exercised by the UK in a way which might have contributed to its 'Dirty Man of Europe' image: namely by a reticence to accept proposals which it considers too costly or which adopt policy approaches that have not been usual in the UK. It has also used some environmental issues to further Single Market objectives and this has not always attracted favourable publicity or acclaim, as the Dutch reaction to the Packaging and Packaging Waste Directive exemplifies. While not generally perceived as a 'leader' in the EU, it has therefore contributed significantly to policy development, although not always in wholly positive ways.

THE IMPACT OF THE EU ON UK WASTE MANAGEMENT POLICY AND ADMINISTRATION

EU policy regarding waste management, all agreed to by the UK in the Council of Ministers, has had significant impacts on the UK. There are various explanations as to why the UK Government has assented to EU legislation: it has at times been seen as a means to introduce domestic legislation which would have proved unpopular with some sections of British economy or society; wider issues have on occasions necessitated negotiating trade-offs; and in some instances, the government may have been unaware of the full implications of the legislation to which it was agreeing. Overlaying all of this are the obligations imposed by having to implement EU legislation, sometimes as interpreted by the European Court, at speeds and levels which otherwise might have been moderated under domestic law. One should be wary, however, of attributing too much causation to the EU for the many changes in UK policy and administration (described in the first section of this chapter) when domestic or other pressures may also have been important. In considering the balance of influences at work it is useful to distinguish between different types of impact – namely, policy substance, policy concepts and policy administration and style.

The impacts of a number of directives in terms of policy substance are readily apparent. The 1978 Hazardous Waste Directive, for example, influenced the way in which the UK developed its own Special Waste Regulations, helping to determine which wastes were to be included and creating a pressure for the latter's rapid adoption (Haigh 1995). The impact of the Regulation on Transfrontier Shipment of Waste has been to hasten the phasing out of imports of waste for incineration and to tighten provisions concerning the export of certain wastes for recovery. More recent policies have also had a discernible effect. The Landfill Directive imposes demanding standards which would probably not have been set in the absence of EU negotiation, even if the UK did achieve a significant exemption over the issue of co-disposal of hazardous and non-hazardous waste. Likewise the Packaging and Packaging Waste Directive has not only now set targets to be achieved but obliged the government to set up systems for administering the Directive which otherwise would probably not have been set up for some years to come. The mere pressure to agree specific targets in the drawing up of the Directive was also a significant driver for the UK to introduce its own (admittedly less comprehensive) targets, so as to influence the negotiations.

The impact of the EU is also discernible in the way in which the UK now considers waste management issues. Probably the clearest example is the repeated debates over the waste management hierarchy, especially the relative priorities accorded to incineration, recycling and re-use. Acceptance of the need to move waste management as far as possible up the hierarchy may in part have been due to domestic pressure, but it is also certainly in keeping with the EU discussions in which the UK has been involved. The priorities which have become apparent in the Commission's waste management strategy and the Fifth Environmental Action Programme were not introduced by UK initiatives. Although the hierarchy is still only 'indicative' in its key provisions, the fact that issues raised at the EU level have assumed a stricter interpretation has undoubtedly affected UK policy thinking. That thinking is moving more and more away from a reliance on landfill towards the mixture of practices adopted on the Continent: more incineration, recycling and re-use.

Another way in which EU policy thinking appears to have had an impact on the UK, at least indirectly, concerns the greater use of economic instruments to achieve waste management objectives. The introduction of the

UK's landfill tax, for example, although not a direct response to EU legislation, can be seen as part of a trend where such measures are advocated in general terms by the EU and are increasingly taken on board in this country. The need to involve all economic and social sections of the production and consumption chain has also received an impetus from the EU, which has itself developed ideas which stemmed more often than not from either Denmark, Germany or the Netherlands. It thus seems that much of the effect of EU membership is through a speeding up of UK policy developments and a more subtle dialectic of 'policy learning' rather than through the linear view of EU policy directly prompting changes in the UK. It is in the discussion of and education about initiatives from other European countries and the subjection of its own practices to international (European) comparison and on occasions criticism and public opprobrium, that most of the key policy changes can be identified.

From the point of view of policy administration, specific procedural changes can be directly attributed to the EU. An example is provided by the way in which the UK has transposed the amended Framework Waste Directive. Whereas with the earlier legislation it took virtually no formal action whatsoever, believing the provisions of existing legislation to suffice, in the case of the amended Directive it has given it almost verbatim transposition in Part II of the Environmental Protection Act (Haigh 1995). This means that the exact definition of waste adopted in the Directive is now applicable in the UK as are the various options in the waste management hierarchy. The reason for this change of approach is the desire to avoid legal proceedings being initiated by the Commission for improper or incomplete transposition (see Chapter 2). The Commission's interest in and responsibility for monitoring the substantive implementation of Directives, as well as their formal transposition, meant that even the original Framework Waste Directive had an effect on the UK, despite apparently being modelled on the Control of Pollution Act. The requirement in the latter that WDAs draw up waste disposal plans was largely ignored until the Commission put pressure on the UK to see that such plans were forthcoming, which it was obliged to do by the Framework Waste Directive.

From the point of view of administrative structures and relationships between the actors involved, there has been some EU influence, but it appears that most change is as much the result of domestic or other pressures. The gradual adoption of 'technology forcing' measures, for example, may be as much because the government has become convinced of the advantages of a 'policy edge' for British waste management companies (Butson 1993) as through its 'imposition' from the EU. The obligation to implement EU legislative commitments has certainly added to the pressures toward centralisation of environmental administration, monitoring, enforcement and standard-setting away from local authorities (Crockers 1993). But other pressures have played their part (see Chapters 4 and 8), and the establishment and form of the Environment Agency cannot be seen as a mere response to this obligation (Carter and Lowe 1995).

However, the simple fact that the EU policy process now gives greater opportunities for input from a wide range of interests, from environmental and consumer groups, to local authorities and business interests, has encouraged a change from the style of policy which emphasised relatively closed and expert consensus to one which is more open and accountable (Chapter 5). The way in which the UK Government has now demonstrated in a public way how it is seeking to achieve the waste management objectives laid out in the Fifth Environmental Action Programme is interesting as much for making UK policy making somewhat more accountable and transparent as it is for the content of the document (UK Government 1994b). Equally, the setting of specific standards will alter the relationship between many of those involved in making and influencing waste management policy.

CONCLUSIONS AND FUTURE DIRECTIONS

This overview of the development of EU and UK waste management policies and practices suggests that there are complex reciprocal influences at work. It is evident that significant changes have occurred to the UK's waste management policies but the extent to which these can be directly attributed to EU membership is somewhat less clear. In some instances, standards are somewhat higher than they would otherwise have been and the obligation to implement legislation agreed in Brussels has contributed to the development of more centrally based waste management structures within the UK, as Haigh (1986) predicted. However, many of the changes to UK policy and policy making are in response to far wider international pressures, as epitomised by the 1992 Rio Conference, as well as to domestic political factors: including greater public awareness of waste management and sustainability issues, industry pressure for harmonisation, party political calculation and recommendations from national committees and UK experts. This is more in line with the conclusions of O'Riordan and Weale (1989) and suggests considerable caution in assessing the relative influence of each factor.

However, some of these more diffuse influences have their origins in developments occurring in other member states which have their own domestic political agendas. Membership of the EU increases the extent to which the UK Government must respond to these different agendas and is brought into contact with different approaches and concepts in waste management. Membership has also increased the potential for interest groups to draw attention to these developments and to use the dependencies created through membership of the EU to elicit responses from the UK Government, raise public awareness of the comparative UK position (be it good or bad) and develop their own positions. The term 'policy learning' (Jachten-fuchs and Huber 1993) characterises such reciprocal influences of the member states, the EU and other political players active in the process.

Looking forward, one can envisage a continued period of change in UK waste management and even the 'quiet revolution' suggested earlier. Much of the impetus for this lies predominantly in UK factors, most notably the way in which the Environment Agency will carry out its work. The greater use of cost-benefit analysis, life-cycle analysis, economic instruments and environmental education, as foreseen by both the Environment Agency and the UK Government's own waste management strategy, all herald further innovation, as do the increasing environmental pressures related to waste management and landfill availability. The relations between the policy makers and the parties affected by legislation are likely to remain unstable: many of the initiatives involved in promoting and realising the concept of 'producer responsibility', for example, are forcing disparate sections of the production and consumption chain to come together to seek common solutions to waste problems which had previously been viewed in isolation. UK policy makers, as well as having to consider the EU and international aspects of their work, will also have to contend with these different and shifting alliances. The problems which the DoE has had in gaining an agreement with the various sections of the UK packaging chain give a good indication of the difficulties ahead. It also exemplifies how the obligation to implement EU legislation quickens the pace at which change is pursued at a national level.

At the EU level, the preoccupation now is more with consolidation and effective implementation than generating lots of new legislation. However, the Commission is certainly devoting attention to the issue of its waste management strategy (EWWE 1995; EIS 1996), the review of which is revisiting some of the key principles of current practice (CEC 1995b). While it appears that the priority waste streams approach has lost some favour (despite ambitious national initiatives in several areas, for example, in Germany), there is renewed emphasis on re-establishing the waste manage-ment hierarchy in such a way that incineration is placed below recycling, recycling placed below re-use and greater attention is devoted to minimisa-tion objectives. Some, but not all of this will sit easily with the UK Govern-ment's current position. The UK might do well to reflect that it is member states with innovative and proactive approaches who tend to set the EU agenda.

213

12

NATURE CONSERVATION

James Dixon

The conservation of biodiversity, the variety of life on Earth, is widely recognised as an activity which benefits the whole of humanity and which requires in turn a shared effort by scientists, administrators, land managers and the public at large (Wynne *et al.* 1994). Many of the values, concepts and practices of nature conservation and much of its global leadership have originated in Western Europe, as have many of the threats to it. UK institutions and individuals have historically played a leading role within both European and global conservation efforts. The nature of this role has changed in relation to a number of factors. UK statutory agencies historically played an important role in European nature conservation, a role which is now played more by central government and the voluntary sector. This chapter analyses this transition and suggests issues to consider for the future.

NATURE CONSERVATION IN THE UK

The UK in context

A number of factors serve to make the UK perspective on nature conservation unique, particularly its physical environment, its popular and scientific appreciation of nature conservation and its policy-making. These are reviewed briefly below to pick out those aspects which shape nature conservation policy.

The UK has a distinctive physical environment, being an island in the Atlantic region with a strong maritime influence and a diverse coastline. It is a crowded and industrialised country, where 80 per cent of the population live in urban areas, with an unusually high proportion of the land surface under agriculture (76 per cent) and a low proportion under forests (9 per cent). Despite intensive land use, small pockets of biological diversity remain in 'semi-natural' habitats (composed of native species but much modified by human practices) in a complex mosaic of rural land use, especially in the wetter, upland areas of the north and west. Post-war land use change has led to a reduction in the stock of habitats and declines in many, once common and widespread species. Through intensive site and species protection, many species have been 'saved' from extinction or even re-introduced (Grimmett and Jones 1989).

The UK has one of the oldest and strongest nature conservation move-ments in the world. It has an extraordinarily large membership (approaching 10 per cent of the population in total) which skews concerns towards cha-rismatic species and habitats (particularly birds, orchids and woodlands, for example) but which has allowed the development of well-resourced and confident voluntary organisations. There is a strong tradition of field natural sciences in universities, research institutes and societies for amateur natural-ists. Both voluntary organisations and government conservation agencies have been very closely associated with this scientific culture. However, the agencies have been poorly funded (in comparison with other departments managing resources of water, forests and agriculture) and not well inte-grated with physical and land-use planning (Nature Conservancy Council 1984).

Policy making in general in the UK has undergone a number of signifi-cant changes since the 1970s. These include: moves to limit government 'intervention'; a more policy-focused civil service; policy implementation by separate agencies of government; the use of market-oriented mech-anisms; and active support of the voluntary sector. In nature conserva-tion, this led to the growth (during the 1980s) of the Nature Conservancy Council (NCC) as the principal 'agent' of government policy, the devel-opment of management agreements to pay landowners to protect and manage sites and wildlife, and support for the expansion of voluntary organisations. Much publicly funded 'science for science's sake' has been replaced by a more customer-oriented approach with agencies and cen-tral government purchasing specific programmes or pieces of applied research from contractors (Winter 1996). This has affected the UK science base for nature conservation. At the same time, changes in sub-national government and international relations have altered the insti-tutional context in which policies for nature conservation are formulated and administered. On the one hand, although the UK is a centralised, unitary state, there has been increasing administrative devolution, espe-cially to Scotland as well as to Wales and to Northern Ireland but also, most recently and tentatively, to the regions of England. On the other hand, the UK's external relations have undergone a transition from man-aging the retreat from Empire to a concern with promoting exports and managing ever-closer relations with a developing European Union. Nature conservation has been affected by these contrasting tendencies, becoming to an extent Europeanised in its policy making and regionalised in its administration.

Historical development of nature conservation

UK nature conservation has existed in three broad and overlapping arenas, namely science, statutory agencies and the voluntary sector. Each of these arenas has influenced the extent to which the UK has engaged in European initiatives. Indeed, the balance between them has changed recently, with considerable consequences for the success of UK initiatives.

In the immediate post-war period, a small but dedicated group of scientists formulated the essential framework of UK nature conservation policy. Julian Huxley and Arthur Tansley's 1947 report on *Conservation of Nature in England and Wales* led to the establishment, in 1949, of the official Nature Conservancy, a strongly science-based organisation with a requirement to 'provide scientific advice', develop a set of research institutions and establish a series of protected areas as 'representative sites'. Although site protection depended on planning controls, nature conservation was poorly integrated with land-use planning. It also had little to do with agriculture, forestry and other natural resource management issues, even though the nature conservation interests of protected sites depended on the continuation of traditional land management practices. No threat was perceived from agricultural change at the time.

In 1973, following concerns that nature conservation should be more to do with action and advice to government than undertaking science, the Nature Conservancy Council was established, shorn of its research institutes. In practice, the NCC was still largely staffed by scientists and it retained a capacity to monitor flora and fauna and to conduct its own scientific surveys and investigations, as well as having a budget to commission research from its former institutes and other organisations. Thus, its development in the 1970s and early 1980s was on a firm scientific foundation. However, the growing politicisation of conservation issues nationally and internationally, the development of agricultural and forestry technologies, and continued economic growth required a more activist conservation agency. In particular, the high rate of loss of semi-natural habitats to land-use change led to the enactment of the Wildlife and Countryside Act in 1981.

The 1981 Act focused the greater part of statutory conservation efforts on surveying, selecting and designating further Sites of Special Scientific Interest (SSSIs) so as to protect them from all threats, not just those regulated through the planning system. To this end, it introduced a mechanism to compensate landowners and farmers for not destroying or for actively managing sites of nature conservation value. Thus, the mid-1980s saw the development of two key concepts in UK, and subsequently, European, conservation. First, UK statutory agencies became expert in the design, administration and evaluation of a protected areas system, and, second, the concept of the management agreement with private landowners was developed. Still, little was done for the conservation of the wider country-side until Agriculture Ministers were given a duty to 'balance' protection of the countryside with agricultural activities in the Agriculture Act 1986.

The growth of the NCC in the 1980s was consistent with central government intentions to devolve the administration of policy to agencies, while retaining policy making centrally. Thus, during the 1980s, the NCC grew dramatically financially and in its public profile, only to be reorganised and much weakened at the end of the decade. Following ministerial irritation at the independence shown by the NCC (particularly over SSSI designation in Scotland), the Environmental Protection Act (1990) split the NCC's functions into separate country agencies for England, Scotland and Wales. Subsequent reorganisations created further devolution of powers to regional offices, teams and boards. The centralised, science-driven agency was superseded by smaller, local teams with less of an expert culture, fragmented influence and a much reduced scientific capacity. There is also little evidence of either the will or resources being allocated to the NCC's successors – English Nature, Scottish Natural Heritage and the Countryside Council for Wales – to allow them to address 'wider countryside' conservation substantially. In addition, this reorganisation will continue to have a significant impact on domestic nature conservation policy for years to come and has also curbed the statutory agencies' role in international nature conservation (see below).

In contrast, the 1990s have seen the voluntary conservation organisations growing in strength, influence and confidence. In doing so, they build on a diverse and long tradition, including, for example, the Royal Society for the Protection of Birds (RSPB) (founded in 1889), the National Trust (in 1895), the Fauna and Flora Preservation Society (1903) and the Wildlife Trusts (1912). Some of these organisations, committed as much to practical action as to influencing opinion, pre-dated the scientific organisations, such as the British Ecological Society (founded in 1911) and British Trust for Ornithology (1933), which developed the new science of ecology. However, throughout, the close association between the two types of organisation has been significant.

Also significant has been the degree to which the voluntary organisations have built up their support in public opinion. In the early part of the century, nature conservation was mainly a pastime of leisured clerics, aristocrats and the genteel middle class. However, that the voluntary organisations now boast a combined membership of nearly 5 million people shows how they have ridden the social changes of this century, successfully developing as a truly mass movement. One result is that the major voluntary organisations are in the fortunate position of having strong financial (and hence political) independence and the ability to develop scientific, technical and policy expertise in a way only possible with government funding in most other countries.

Key characteristics of UK nature conservation

Historically, the strategy adopted by conservationists was species-driven. The middle decades of this century added site acquisition and management. The 1970s and 1980s added the protection of networks of sites and the regulation of land management practices. Finally, the 1990s have identified the need to alter the policies that drive land-use change. Table 12.1 identifies the complementary approaches now adopted. However, although this strategy appears comprehensive, it is delivered by different agencies and organisations and risks a lack of overall co-ordination and integration. Moreover, the complexity and difficulty in achieving the 'processes' of nature conservation have often entailed (in a classic means–ends displacement) losing sight of the fundamental goals. The Biodiversity Challenge (Wynne *et al.* 1994) initiative, promoted by the voluntary conservation organisations and the Government's Biodiversity Action Plan (UK Government 1994a), developed in response, aim to focus attention on the achievement of specific conservation goals by applying the strategy as a whole. This creates yet another opportunity to address the conservation of the wider countryside, but it is too early to see whether this will be successful.

The *strengths* of UK nature conservation lie in its strong science and voluntary base; its internationally recognised high standards of nature reserve management and conservation practice; a proven and advanced site and species evaluation methodology for setting conservation priorities; strong compliance with international legislation; and a high level of public support. Its *weaknesses* are that its science base and statutory sector are in relative decline; there is poor integration with planning and land use; and, most significantly, there has been repeated failure to address the conservation of the wider countryside.

Table 12.1 Elements of a strategy for nature conservation in the UK

- Scientific rationale for action, research and monitoring
- Species protection and recovery programmes
- Habitat protection strategies
- Site identification, evaluation and protection
- Reserve purchase, management and multiple use
- Strict control over change of use for special sites
- Management agreements and voluntary incentives for 'sub-SSSI' or larger-scale sites
- A mixture of incentives, regulation and economic reform in 'mainstream' land uses such as agriculture, water and forestry
- Sustainable development strategies at national and local levels
- Education, awareness and public support gathering

Source: Wynne *et al.* 1994

UK NATURE CONSERVATION IN EUROPE

The growing significance of the European context

The UK has 'exported' much of its expertise in nature conservation, has engaged in international fora with enthusiasm and has influenced the international conservation agenda. It has also adopted practices and strategies reflecting international obligations. A significant focus of this effort has been on European nature conservation. There are a number of reasons for this. First, in order to influence UK legislation it is increasingly necessary to influence EU legislation. Second, many pan-European legislative structures and strategies exist in Europe. Third, transboundary action is necessary, for example, for migratory birds and large-scale site protection. Fourth, the only way of addressing international economic sectors (such as agriculture) is at the appropriate international level. As in the development of UK institutions and policy, a complex and changing blend of science and statutory and voluntary activities has represented the UK in international conservation.

Science

Traditionally, the natural sciences have been international in outlook and, with a growing awareness of international conservation problems and developing communication technologies, this is ever more the case. Examples of where UK conservation science has been internationally significant include the historical ecology of woodlands (Peterken 1996); the 'unravelling' of the link between organochlorine pesticide use and decline in bird populations (Moore 1987); and the understanding of coastal processes.

Today, the UK hosts a number of international and national voluntary organisations and institutes which contribute substantially to UK and global conservation science. These include the centres of excellence in taxonomic studies, such as the Royal Botanic Gardens, Kew, the Natural History Museum and several university departments. There are also major international inventories of biological data: on biodiversity, at the World Conservation Monitoring Centre (Cambridge); on European bird populations, held by BirdLife International (Cambridge); on wildfowl, at the International Waterfowl and Wetlands Research Bureau (Slimbridge); and on freshwater fish at the Freshwater Biological Institute (Windermere).

In a very modest way, the UK Government actively supports the development of international databases by the transfer of expertise and financial support to initiatives such as the European Environment Agency (see Chapter 7). Bilateral aid is also provided through both the Darwin Initiative (globally) and the Environment Know-how Fund for Central and Eastern Europe, for biodiversity initiatives. However, these are small in comparison

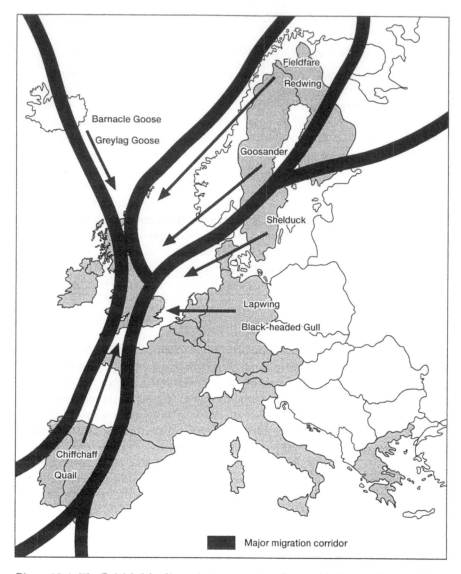

Figure 12.1 The British Isles lie at the intersection of major bird migration corridors. Examples here are of migrants to Britain from other parts of Europe (though most come from further afield). Winter visitors hail from the north and east; summer visitors from the south-east and south-west. The primary motivation behind the EU's Birds Directive was to protect migrating birds.

with other environmental priorities and in comparison with that provided by voluntary organisations. A number of UK research institutions have developed close links with related institutions overseas, although less often in Europe than in developing countries. The availability of EU research funds (which require collaboration between institutes in different member states) has helped overcome this.

Statutory agencies

The changing international role played by UK statutory conservation agencies is probably the most notable development in the UK's evolving relationship with European nature conservation. Historically, the science-led agencies maintained extensive scientific links internationally. The Nature Conservancy's annual report for 1962 includes eight pages of international activities, including scientific exchanges, attendance at conferences and hosting international scientific gatherings, and participating in meetings of international voluntary organisations, such as the International Council for Bird Preservation (now BirdLife International), the International Wildfowl Research Bureau (IWRB), and an active role in the World Wildlife Fund (WWF) (which Nature Conservancy staff had helped to establish). Staff were likewise playing a catalytic role in establishing the Council of Europe work on nature conservation and taking a lead in the activities of the International Union for Conservation of Nature and Natural Resources (IUCN). Biographies of the most influential people involved – such as Max Nicholson (1970) and Norman Moore (1987) – illustrate clearly that individuals from the statutory agencies led the development of international initiatives. However, the UK conservation agencies no longer play this role to such a degree.

In the 1960s, the statutory conservation agencies contributed ideas, influence and resources to international conservation initiatives. However, with the UK accession to the (then) EEC, and the conclusion of a number of non-EU international treaties (such as the Bern Convention on European Wildlife and Natural Habitats), the statutory role became a more formal one. With more and more at stake, central government came to feel that it could not entrust such international negotiations to scientists from a semi-autonomous agency.

International nature conservation has become a matter for central government, particularly the Wildlife Branch of the DoE (see Chapter 2). Career civil servants (including some seconded from the agencies) now take the responsibility for negotiations on international legislation and its compliance. Inevitably, the levels of personal expertise and commitment are less than was found within the specialist agencies. Moreover, government looks to DoE civil servants to achieve international legislation for nature conservation constrained by other factors such as the cost implications for UK business and consistency with overall government policy, especially policy towards Europe (see Chapter 2).

Devolution of conservation responsibilities to the countries of the UK, and increasingly to regions, and the relative scarcity of centres of expertise, mean that the agencies have less to offer an international audience. A degree of introversion has set in in the statutory sector, with the partial exception of the much battered and resource-starved Joint Nature Conservation Committee (which was set up to co-ordinate the country agencies following the NCC's demise). The JNCC has tried valiantly to develop a programme of international conservation, notably its European Pastoralism work (McCracken and Bignal 1995), only to have its programme subject to scientific and other criticisms from the home agencies.

Reflecting this introversion, UK agencies and central government play only a small role in supporting nature conservation overseas. For example, of the £4.9 million allocated by the Environment Know-how Fund (to countries in Central and Eastern Europe) between 1992 and 1995, only 4 out of 120 projects (amounting to £34,900 or 0.3 per cent) were allocated to nature conservation (House of Commons Environment Committee 1995). This contrasts with the substantial sums allocated to European nature conservation by the Swedish, Swiss and Dutch nature conservation authorities.

The voluntary sector

In contrast, UK voluntary organisations have taken a lead in international nature conservation. Several organisations, notably WWF, RSPB and BirdLife International, Flora and Fauna International and IWRB, have significant international programmes, aimed at 'global' priorities, for example, in the tropics, off-shore islands, and the Antarctic. They also have important European programmes.

Activities supported, for example, by the RSPB in Europe include:

- survey and data collection on a pan-European scale (such as the Dispersed Species Programme) (Tucker and Heath 1994);
- development of Europe-wide conservation programmes (such as for Protected Areas, or the recovery of globally threatened species);
- assistance to foreign voluntary organisations with capacity building (especially membership development, core funding, staff training and development);
- support to site protection and species campaigns in other countries;
- joint action on European policy and legislation (in fields such as agriculture, transport, the Structural Funds).

The growth in international work by voluntary organisations has reflected both an increased capacity to do so and the increasing internationalisation of conservation work (see Chapter 6). The RSPB supports the operation of BirdLife International and its partners across Europe to the tune of nearly £1 million per year. WWF-UK is one of the largest funders of the WWF

International programme. As UK voluntary organisations develop, these contributions are planned to increase, although difficulties can arise if members begin to feel too great a proportion of funds is being spent over-seas and UK charitable fundraising legislation limits the extent to which funds can be 'handed', without close scrutiny, to international projects. Therefore, while voluntary organisations have a key role to play, they can never hope to provide the level of funding that statutory agencies and central government should.

THE DEVELOPMENT OF EUROPEAN POLICY

Legislative frameworks

Early European legislation for nature conservation was determined under the auspices of the Council of Europe, notably the Bern Convention on the Conservation of European Wildlife and Natural Habitats (drawn up in 1979) and the European Network of Biogenetic Reserves (1976). Specific agreements were developed for migratory birds and mammals under the Bonn Convention (drawn up in 1979). Both the UK and other European countries are signatories to global legislation or programmes such as the World Conservation Strategy (1980), Man and Biosphere Programme (launched in 1970) and the Convention on Biological Diversity (agreed 1992). Overall, the UK has proven a responsible signatory of these pro-grammes but their impact has been relatively limited in that, mostly, these conventions relate to a very small number of sites and species. Of greater importance have been the policies of what is now the European Union (see Table 12.2).

Prior to the UK's accession to the EC in 1973, nature conservation fea-tured little in Community legislation. Those member states with an interest originally maintained that nature conservation was not a subject for Com-munity competence. Consequently, the first Environmental Action Pro-gramme achieved little for nature conservation, except by proposing a study on migratory bird protection. This subsequently led to the adoption by the Council of the Directive on the Conservation of Wild Birds during the Second Environmental Action Programme in 1979.

This first substantial piece of EU legislation was significant for two reasons. First, it was an early indication of the potential role of voluntary organisations in developing EU legislation: the RSPB, and its international partners played a key role. The 'Birds Directive' was also significant because it set an important precedent for a more all-embracing (but politically more difficult) EU Directive on the Conservation of Natural Habitats and of Wild Fauna and Flora (the 'Habitats and Species Directive') which was not agreed by the Council of Ministers until 1992.

Table 12.2 Key nature conservation legislation and agreements since 1973

UK legislation	EU legislation	Council of Europe	Global agreements
			CITES 1973 (Convention on International Trade in Endangered Species)
	Birds Directive 1979	Bern Convention 1979 (Conservation of European Wildlife and Natural Habitats)	Bonn Convention 1979 (Conservation of Migratory Species of Wild Animals)
Wildlife and Countryside Act 1981	Article 19 (ESAs) of the 1985 Agricultural Structures Regulation		World Conservation Strategy 1980
Agriculture Act 1986			
Environmental Protection Act 1990	Agri-Environment Regulation 1992 Habitats and Species Directive 1992	Environmental Action Plan for Central and Eastern Europe 1995 European Biodiversity Strategy 1996	Bio-diversity Convention 1992

The UK voluntary organisations actively campaigned for the Habitats and Species Directive but it was opposed by those countries where it would have the greatest economic impact, particularly Spain (see Chapter 2). The Commission finally managed to achieve sufficient support only after agreeing two budget lines to pay for its implementation, one directly tied to the Directive (under LIFE) and the other, much more substantial element under the EU's Agri-environment Regulation (EEC 2078/92) (CEC 1992a) – an indicator of the growing importance of land use, agriculture and 'wider countryside' concerns for nature conservation (BirdLife International, WWF and World Conservation Union 1995).

More recent EU initiatives (under the Fifth Environmental Action Programme, for example) have concentrated on an attempt to integrate nature conservation and environmental considerations into other policies. Little progress has been made with new conservation legislation in the EU as member states are now preoccupied with implementing the Habitats and Species Directive and, indeed, curtailing the implications of some of the earlier legislation. Of more significance now is a programme of strategic development of environmental policy for the whole of Europe involving a number of international organisations – the EU, Council of Europe, World Bank and the UN Economic Commission for Europe.

Following the collapse of the former communist regimes in Central and Eastern Europe, all western agencies have begun programmes of political association with these countries. For the environment, this has been under the auspices of the 'Environment for Europe' process. A number of result-ant documents, including the *Environmental Action Plan for Central and Eastern Europe* (Council of Europe 1995b) and the *Biodiversity Strategy for Europe* (Council of Europe 1995a), were endorsed at the meeting of Europe's environment Ministers in Sofia in October 1995. These docu-ments set a strategy for bilateral and multilateral aid (especially from the EU, World Bank and Global Environment Facility) and for governments in Central and Eastern Europe to develop their institutions. The UK ought to be playing a leading role in the implementation of the strategy, but it is not.

Land use and economic policy

A recurrent theme of this chapter is that nature conservation policy makers in the UK have often failed to acknowledge the importance of wider land-use issues. This has been partly because of the enormity of the task. The causes of conservation problems (threats to species, habitats and sites) are complex and the result of changing land uses, especially farming. Increas-ingly, also, the problems originate in and the solutions reside in European policies. Action at a European level requires political will, considerable resources, expertise and the capacity to build alliances to influence EU legis-lation. Rarely, until recently, were conservationists able or willing to commit these. Now, however, it is becoming difficult to avoid doing so in recogni-tion of the greater role that European institutions have come to play in the government of the UK.

Closer European integration has resulted in additional economic pres-sures on the UK environment. Sectors such as port development, motorway and railway construction, fisheries and rural development have changed dramatically as the UK has sought to maximise trade benefits within the Single Market. Wider economic trends and policies, such as the potential impacts of Economic and Monetary Union, enlargement of the EU to embrace the Central and Eastern countries and the international 'footprint' of the EU (the environmental and resource impact of its global trade rela-tions) have been subject to little or no detailed analysis or concerted lobby-ing from UK conservation organisations. The more tangible policy area of agriculture has received most attention and so is considered here as a case study.

Greening the Common Agricultural Policy

Early concerns surrounded hedgerow and woodland loss but arterial drainage and conversion of habitats such as heathland, moorland and grassland to cultivated land or rye-grass monoculture happened at an unprecedented pace. While many of these landscape changes were directly encouraged by UK policies (especially provision of advisory and research services, the favourable tax and planning status of farmers and capital grants) it was the prices and subsidies received by farmers that drove the pace of change, and these were largely set under the CAP.

Up until the mid-1980s, a few siren voices in conservation and government (and a very few farmers) pointed out that this could not last. It was argued that EU markets would rapidly be satiated and prospects for exporting at world prices were small without unreasonable levels of subsidy. Critics also pointed to the damage to the countryside caused by the CAP and to the high costs to consumers and the food processing industry. But in the 1970s few in the UK Government or the EU saw the need to change the CAP (Winter 1996).

UK environmentalists were slow to point the finger at the Common Agricultural Policy as the cause (and the solution) to wildlife and landscape change. Efforts were directed at minimising, if not stopping, the more obvious examples of landscape change in, for example, the river valleys of England in the mid-1970s, the coastal grazing marshes of the south and east of England throughout the 1970s and 1980s and the uplands, especially in Wales, in the early 1980s. That the Common Agricultural Policy was 'responsible' became evident to a small number of individuals in the major conservation organisations (such as the statutory agencies, CPRE, FoE and the RSPB). However, corporate action by these organisations was limited for a number of reasons, notably the great cost of doing anything about it, the political difficulty of confronting a strong and unified farming industry and preoccupations with other legislative issues.

Riding the beginning of a wave of criticism of the Common Agricultural Policy in the early 1980s, the leading voluntary organisations in particular began to consider campaigns to reduce the damaging consequences of the CAP. In parallel, the statutory agencies, while shrewdly avoiding confrontation with MAFF and the farming industry, began to trial practical ways of 'integrating' farming and conservation. This was initially through advice and friendly coercion, but the Broads Grazing Marsh Scheme broke new ground – or rather sought to prevent this!

In the aftermath of the disputes surrounding designations under the Wildlife and Countryside Act (such as the Berwyn Mountains and the Somerset Levels) attempts were begun to broker a compromise solution, whereby farmers were paid not to damage the countryside through a management agreement (Lowe *et al.* 1986). This concept, previously developed in parts of the Netherlands and simultaneously developed in Germany and the Denmark, was the forerunner of the Environmentally Sensitive Areas and the EU's Agri-environment Regulation. Its development, through the first and second round of ESAs in England, Scotland and Wales, was greater in the UK between 1987 and 1992 than in any other EU member state. Indeed, during the run-up to the 1992 CAP reforms it was the UK agriculture Departments who truly championed this approach. Sadly, this pre-eminent position has not been maintained subsequently (BirdLife International 1996).

In the management agreement concept, conservation organisations saw a resolution to the gathering difficulties caused by agriculture. In a number of fora and publications during the mid-1980s, the 'UK view' on the CAP began to develop. Political, consumer and environmental perspectives began to coalesce as the costs of the CAP in budgetary and environmental terms escalated. Poor world prices, surplus production and a resistance to change within the EU institutions led to successive EU budgetary crises during the period 1984 to 1988. A strong 'free market' ideology under Margaret Thatcher's leadership and a growing climate of 'Euroscepticism' provided fertile ground for criticisms of the CAP. During this period, an increasing number of conservation organisations in the UK used their growing resources to develop a more substantial capacity to seek changes to the CAP. Three voluntary organisations played a leading role, namely WWF, CPRE and the RSPB. These were by no means the only environmental voices, but they combined independence from government, the ability to employ specialists in agricultural policy and economics, and strengthening networks across Europe. Earlier critiques of the CAP suffered from being idealistic, poorly addressed at other European policy agendas, and weak on solutions. These organisations, in contrast, could bring a growing sophistication to bear in their representations on both agricultural and European policy matters.

It would be wrong to suggest that voluntary organisations in other member states were not concerned, rather they were constrained by their own national perspectives (generally pro-CAP in comparison with those of the UK) and were poorly resourced in comparison with the UK conservation movement. EU-wide structures, particularly the European Environmental Bureau (EEB), were a meeting place for like-minded environmental groups, but proved ill-suited for representing this interest with a common, credible voice (see Chapter 6). In the debate prior to the preparation and subsequent conclusion of the MacSharry CAP reforms, UK groups (operating largely outside the EEB framework) played a significant role in encouraging an environmental element to be incorporated into these reforms. This influence was through raised technical understanding of environmental issues within the European Commission, carefully focused lobbying of these officials by 'reasonable' voluntary organisations, and support in the

Council of Ministers for environmental issues from the, then, British Agriculture Minister John Gummer.

Subsequently, UK voluntary organisations have developed a degree of influence over CAP policy making that would have been unimaginable to environmental lobbyists ten years ago. This is for several reasons. The CAP now has a significant (if not central) environment programme which must be administered effectively and the involvement of voluntary organisations has been sought in this. The voluntary organisations have developed further their technical capacity to analyse and propose solutions to *agricultural policy* problems, including those relating to rural development and the new rural policy agenda. Most importantly, the voluntary organisations now have a number of more robust EU-wide networks (notably the BirdLife International partnership and the WWF network), besides the EEB, where common positions can be developed and a voice more representative of the EU as a whole can be articulated.

Carefully timed interventions made by voluntary organisations have been instrumental in achieving positive changes to the administration of the Agri-environment Regulation (such as the requirement to have environmental monitoring) and the EU's strategy paper on enlargement and agriculture (which includes a strong emphasis on environmental measures) (CEC 1995e). While few of these achievements could be made without sympathetic Commission and member state officials and political figures, the voluntary organisations have clearly played a more effective role than previously in agricultural policy. There now exists an informal 'network of networks' (bringing together BirdLife International, WWF and EEB). While this is representative of a wide range of countries, UK organisations have a disproportionate role in each network.

Although voluntary organisations have an increased role, their influence over most CAP decision making is still modest for a number of reasons. Much of the CAP is not primarily 'environmental' and decisions inevitably have a large producer, trade and agricultural administration element. While better organised, the resources of the voluntary organisations are still small and not fully united. Their influence within national agriculture Ministries is very variable, being strong in the Netherlands, Spain, Sweden and the UK, moderate in Denmark, France and Austria, and weak elsewhere. Many CAP decisions are taken in the secrecy of the Council, the Special Committee for Agriculture and the many commodity management committees. The Agricultural Directorate of the Commission has thirty advisory committees for liaison with interest groups and, as yet, there are no environmental representatives: instead support is largely for farmer-representatives, with some co-operative and consumer interests also represented.

The lack of substantial effort on a similar scale from the statutory agencies is noticeable. While they have toyed with agricultural policy, even venturing into some EU-wide policy analysis, this has been piecemeal and with restricted resources and access to European policy makers.

INFLUENCES OF EUROPEAN INTEGRATION ON THE UK

The preceding sections have considered the impacts that UK institutions have had on European environmental and nature conservation policy. However, there have also been impacts on the policy and practice of nature conservation domestically as a result of integration. These have been due to economic impacts, requirements to implement legislation, and the presence of a supranational tier of government able to influence – and challenge – national government and agencies. A 'Europe-wide' perspective has also helped focus conservation efforts on species and habitats considered important in a European, and not just UK, sense. This latter influence is felt strongly in the development of national Red Data lists, notably the revised Red List for birds (RSBP *et al.* 1996).

UK conservation agencies – acting on behalf of government Ministers who agree to EU legislation – have undertaken an exhaustive programme of designation of Special Protection Areas under the Birds Directive and, more recently, the identification and designation of Special Areas of Conservation under the Habitats and Species Directive. Generally, the practical policy mechanisms used to do this have been those under the existing Wildlife and Countryside Act. Indeed, the relative ease with which the UK has been able to implement these Directives is in large part due to the existence of the data and management procedures necessary for that Act. However, the designation of sites under EU legislation confers an extra degree of legal protection and additional requirements, for example, that sites should have management plans. Funding is also available for management work under the LIFE scheme. The greatest impact of these European designations is that there is now a supranational tier of government, to which disputes can be taken. The UK government has itself been challenged over decisions on sites – notably Cardiff Bay and Lappel Bank – and this provides an effective foil to decisions taken solely at a national level.

FUTURE PROSPECTS

New legislation on nature conservation in Europe is unlikely under the climate of subsidiarity which has reduced the overall level of EU legislative proposals. However, there is likely to be a greater emphasis put on 'partnership' approaches between the member states and the Commission in the

Plate 12.1 The Government, for commercial reasons, omitted Lappel Bank from the designation of the Medway Estuary as a Special Protection Area under the Birds Directive. The Royal Society for the Protection of Birds complained to the European Court of Justice, which ruled in 1996 that the UK Government had acted wrongly. Unfortunately, by then Lappel Bank had been developed as a storage area for cars.

Source: RSPB/Gomersall

implementation and elaboration of existing legislation. This will include further development of funding mechanisms such as LIFE.

The greater prospect for furthering nature conservation in Europe is through the better integration of conservation objectives into other EU policy areas. This is an aspiration of the Fifth Environmental Action Programme. However, there are undoubted difficulties to be overcome. Not the least of these is the institutional difficulties faced by environmental professionals attempting to influence economic sectors. This chapter has described how British environmental organisations – most recently dominated by the voluntary groups – have been able to make some progress with greening agriculture policy. Environmental organisations have similarly influenced transport, rural development, fisheries and trade policies. What is clear, however, is that the resources and expertise required to understand such economic sectors are considerable and in short supply.

Moreover, a combined effort by research, statutory and voluntary organisations is necessary. And this effort must be fully embedded within European thinking and networks. Undoubtedly, key organisations will need to learn anew to participate and take a lead in European-wide initiatives. It is vital to recognise as fact the role of EU institutions in the government of the UK. We are not, in practice, faced with a choice between engaging or not. The economic developments – ports, roads, agricultural and rural development – that will come with whatever future model of European integration is chosen, will all impact on wildlife and wild places. Nature conservationists will continue to be confronted with the environmental implications and will need to establish ways of minimising or alleviating these impacts.

13

ENVIRONMENTAL PLANNING

Land-use and landscape policy

Fiona Reynolds

Questions of land use and landscape are arguably very domestic, local pre-occupations. Decisions which directly affect them are made very close to the ground – the way a hedgerow is managed, what kind of development gets consent, for example – and debates about land use and landscape often reflect particular cultural values, traditions and local perceptions.

The UK has strong traditions and perspectives in both land use (exemplified in the relatively early and distinctive development of the UK's land-use planning mechanisms) and landscape (which, almost uniquely in Europe, is separated from mechanisms designed to address the needs of wildlife). This distinctive perspective, strongly felt and even more strongly argued about, helps explain why these issues have a particular resonance in the UK and shapes much of the analysis in this chapter.

Although there are many dimensions of policy that could be caught under the heading of land-use and landscape policy, this chapter looks in detail at only three: agriculture (dominating the vast majority of the undeveloped land area of all fifteen EU member states); 'landscape' policy itself, in its emergent form and as it relates to other key initiatives; and the planning dimensions of land-use policy. The first of these has long been dominated by an EU policy framework, whereas the second and third have until recently been seen as of domestic concern alone, although this chapter argues that this is changing fast.

There are many people and organisations interested in these issues, not to mention the large and often unfocused popular interest stemming from high levels of public concern about how land is managed. Many of these interests, however – the farming community, with its long-standing access to decision makers, and the environmental groups – are primarily motivated by local issues that only infrequently and indirectly get passed up the decision-making chain to the national or European level.

Where policy is actually made and at what level to intervene most effectively on any particular issue pose recurring challenges for all institutions, whether from the official, private or voluntary sector. Experts in lobbying at the national level can be daunted by the complexity of European decision making, and may not even try systematically to influence it. Others find, sometimes by accident, that European policy makers can be easier to influence and get access to than domestic civil servants, and there have been some stunning successes in influencing policy at the European level as a result.

It is important to remember, however, that the range and number of policy influences affecting the land are huge, and they emanate from both the EU and national governments, as well as being very susceptible to differences in practical implementation on the ground. Where, who and how to influence policy change presents those interested in the land with a considerable challenge.

AGRICULTURE

The most profound influence on land use and landscape is undoubtedly agricultural policy. The impact of nearly forty years of production-oriented price support has been subjected to extensive comment and analysis (Bowers and Cheshire 1983; Nature Conservancy Council 1984). Perhaps the most important messages to emerge relate to the adverse environmental impacts of three broad trends: intensification, specialisation and homogenisation of agricultural practice. These trends have changed the face of the British countryside and that of other member states, though often to a lesser extent. Where locally distinctive land uses and activities survive – for example, particular styles of hedge laying, unusual breeds of farm animals, the retention in use of traditional farm buildings – it has more often been in spite rather than because of the influences of agricultural policy.

Only very recently – through, for example, the Environmentally Sensitive Areas scheme, Countryside Stewardship and emerging consumer markets for locally distinctive or environmentally labelled products – has there emerged any kind of structured support, however small, for land-use activities and patterns that are not governed by conventional economic objectives. These reforms were won, largely on the initiative of environmental groups, after bitter battles in the UK over the adverse environmental impacts of agricultural policy (Lowe et al. 1986). Because of the dominance of European policy making in agriculture, to win changes to policy environmentalists had to persuade reluctant Agriculture Ministers to seek new European powers to support environmentally friendly farming, which could then be implemented in the UK. Such support, although enlarged under the 1992 reforms, still has little influence against the overwhelming bulk of Common Agricultural Policy (CAP) payments. In 1996 environmental support constituted only 2 per cent of the CAP budget. But politically they are very popular, and with Agriculture Ministers throughout Europe increasingly relying on

Figure 13.1 Environmentally Sensitive Areas (ESAs) in the UK. These are supported under the EU's Agri-environment Regulation (2078/92). The UK has taken a lead in seeking to 'green' the Common Agricultural Policy.

environmental and social reasons to justify the huge overall expenditure, they may hold a clue to at least part of the CAP's future development.

The UK Government's position on the CAP has changed significantly since joining the EC in 1973. In the entry negotiations the CAP had been amended significantly to reflect UK farming interests and it maintained the high levels of production support to which the UK Government was already committed. During the 1970s and 1980s, the UK Government both argued in Brussels for the retention and development of a production-oriented CAP and implemented its policies in the UK in ways which had dramatic environmental impacts. The policy was undoubtedly successful in its own terms, as production levels and the intensification of farming increased dramatically. But this was at great and evident cost to the environment, to which the hostility between the farming and environmental interests at that time bears testimony.

Not least because of the intensity of domestic pressure, particularly in the early 1980s from environmental groups, the UK Government was one of the first to promote an environmental dimension to the CAP, and successfully negotiated for the establishment of legal powers to support environmentally friendly farming in certain areas (Environmentally Sensitive Areas) in 1985. The success of this instrument, particularly in public relations terms, led to the UK Government incorporating a limited environmental component in its official negotiating position on the 1992 CAP reforms. But wider concerns about the cost of the CAP, expressed especially by the Treasury, have been more dominant. The UK Government's current view that production supports should be withdrawn, in order that agriculture should operate closer to the market, reflects this economic perspective. Environmental and social concerns, while present, are clearly subsidiary in the Government's view to a market-driven, free-trade perspective.

Land-use policy under the agricultural support system has seen a period of relative stability since the 1992 CAP reforms, but is now entering another intense period of debate about its future. Further reform is inevitable. External influences include the reopening in 1999 of the WTO talks on agricultural trade, and EU enlargement eastwards. Internal influences include the cost of the CAP (notwithstanding the current budgetary surplus); its unpopularity, especially with consumers; and growing tensions, sharpened since the accession of Austria, Sweden and Finland, between member states with different views on the role of agricultural policy.

Views about the possible or even likely outcome of the next round of CAP reform are many. However, Franz Fischler, the Agricultural Commissioner, provided a useful benchmark in a paper published in December 1995 setting out three options for reform (CEC 1995e). These can be characterised as: (1) the status quo; (2) the 'radical economic option'; and (3) further reforms along the lines of the 1992 CAP package (that is, a moderate decline in production support and the introduction of a formal, though

Plate 13.1 Opposition to farmers intending to drain Halvergate Marshes in East Anglia led the Ministry of Agriculture to seek European approval for payments to support conservation-oriented farming in 'environmentally sensitive areas'. This was the beginning of EC agri-environment policy.
Source: CPRE/Trelawny

modest, commitment to environmental support from farmers). Commissioner Fischler favoured the last of these. The second is however likely to be given a strong run with the backing of the USA and the Cairns Group.[1] It was also the option favoured – without detailed analysis of the environmental or social consequences – by the UK Ministry of Agriculture's 1995 CAP Review Group whose approach both the then and new Government have broadly endorsed. In the run-up to the reopening of the CAP negotiations, Fischler made a bold attempt to place rural development on the agenda, at a conference in Cork in November 1996. While the Cork Declaration has not been formally endorsed, it achieved a new and broader focus for discussions of rural policy generally (CEC 1996f).

Some of the issues for the next CAP review are beginning to emerge. These include simplification (both bureaucratically and to favour greater subsidiarity); decoupling of agricultural support from production to a greater or lesser extent; and the emergence of environmental and social objectives (in different mixes and with different products sought in different member states) as a key justification for public intervention and support for agriculture (CEC 1996f).

Viewed optimistically, these trends could lead to a significant increase in the resources available for supporting environmentally friendly farming and the delegation of decisions about precisely how such supports should be directed to a much lower level, allowing sensitivity in the application of funds. This is something environmental groups have been demanding for some time. MAFF's establishment of national and regional steering groups in 1996 to guide agri-environment spending is a welcome if small step in this direction. Such changes could bring real benefits for the landscape, wildlife and countryside. But more market-driven farming systems could also have perverse and unpredictable environmental and social effects – for example, encouraging ranching across the hills and uplands; high-intensity cereal farming, with little recognition of the environmental consequences, where it is profitable to do so; and specialist, large-scale livestock units requiring buildings and associated infrastructure.

The net environmental impact of these and other policy changes is therefore difficult to predict. Overall, it is unlikely that CAP reform would significantly reverse the trends towards intensification, specialisation and homogenisation which have been a feature of UK agriculture since 1974: indeed, a market-driven CAP reform could well exacerbate them.

LANDSCAPE POLICY

There is no 'landscape policy' as such at EU level, nor does this seem likely in a formal sense, not least because of the lack of a Europe-wide constituency for the concept of 'landscape'. However, there are many instruments of policy (apart from the CAP) that have a direct bearing on landscape quality and character, and their evolution is likely to reflect, more explicitly in the future than in the past, the qualitative and cultural concerns that are central to the definition of landscape.

A crucial source of influence is Europe's wildlife legislation, stemming from that most mobile of species, the migrating bird. Initial legislation (starting with the Birds Directive (79/409)) was based on the simple fact that protecting a species in one county achieves nothing if it can be shot, persecuted or otherwise damaged in another. Other European legislation followed, of which the most recent is the Habitats Directive (92/43) (see Chapter 12). This recognises explicitly the requirements of both individual species (which need to be protected wherever they are) and the habitats that support them. It argues for a network of connecting sites in order to maintain vulnerable species in a favourable conservation status. In other words, it provides a positive protective framework, not just a means of holding the line against extinction. It also formalises the broader context within which habitats and species survive and flourish, thus moving away from the narrow, site-based approach that characterised much earlier nature conservation policies at both member state and European level. For the first time, individual habitats are seen as part of a continuum in which the rare and the commonplace, and thus the landscape itself, are all a part.

At one stage during the negotiation of the Habitats Directive there was a proposed Annex on landscape features (field boundaries, small areas of woodland, water bodies, and so on) for which it was intended there would be specific requirements for their retention and management. The Annex was drafted by the Commission and inserted in the draft Directive because it was recognised that landscape features constitute important reservoirs, often of a linear nature, for wildlife, and that safeguarding them would aid the fulfilment of the Directive's aim to protect species in their wider context. However, the Annex was deleted and the Directive considerably simplified after the parliamentary scrutiny process, which revealed two contradictory pressures: those from member states like the UK which wanted to limit the detail and bureaucracy associated with the Habitats Directive; and those like Spain and Portugal which were anxious to reduce the costs and burden of implementation.

While many EU member states have designated National Parks and provide other means of identifying and protecting culturally important landscapes and landscape features, it has been left to institutions other than the EU (for example, the Council of Europe, the IUCN) to establish what is being done in different countries and review whether a more coherent approach would be beneficial. Some member states have formal legal bases on which to draw: for example, the Portuguese Constitution has a reference to the role of the state to 'protect and enhance the cultural heritage of the Portuguese people, protect nature and the environment and preserve natural resources'; others do not separate landscape from wildlife protection at all. A wide variety of mechanisms embracing legal safeguards of varying degrees of strictness, positive incentives, voluntary schemes and mandatory requirements is used throughout the member states. While proud (though only partly justifiably, given evidence of serious and continuing decline in the landscape quality and character of the British countryside) of the UK system, the British Government has not initiated debate about possible EU dimensions of landscape policy. This is partly from a reluctance to expand the European basis of legislation in general; and partly because the landscape is genuinely seen as a domestic matter, and not something which needs European legislation to protect it.

The Maastricht Treaty went a step towards formally recognising the concept of cultural resources, which arguably include landscape, in the new Article 28 of the Treaty, which refers to the cultural heritage of member states and the Union. This raises a fundamental question about what (if anything) constitutes a distinctively European heritage, while also – in a practical way – encouraging member states to better document and value their own cultural heritage. There is still little certainty about what this new

Article might mean in practice and naturally there are different perspectives on what cultural resources mean. In the general climate of anxiety about over-zealous action at the European level, it seems likely that the first steps will focus on the more easily defined elements of our cultural heritage (the built environment, for example) and action that can be taken at member state level to help better define it, share knowledge and experience and consider whether there is a need to establish common principles for its protection. The UK Government, as often before, is likely to argue that its existing domestic legislation provides an adequate response to such suggestions. But it is possible that what will emerge from these discussions is a more clearly articulated sense of what it is to be European (recognising, of course, the artificiality of the current EU boundary for this purpose) and how our cultural landscapes and artefacts contribute to a sense of collective identity.

LAND-USE PLANNING POLICY

This term means a wide variety of things, especially in a European context. This chapter selectively groups three elements: the formal land-use planning system (usually defined by planning legislation, guidance and policy in individual member states); environmental instruments which use planning or analogous procedures of review and consent (especially Environmental Impact Assessment); and the growing need to reduce conflicts between strategic land-use goals and EU funding instruments. There are interesting developments in each of these areas, although their long-term outcome is far from clear.

The land-use planning system

Land-use planning *per se* has long been held emphatically to be a member state responsibility, notwithstanding the fact that many countries face common land-use challenges and have to deal with much common European policy. Widely differing land-use planning regimes operate in different countries, and there has been little attempt until very recently even to explore whether a more unified system would bring benefits. However, in the 'Europe 2000' exercise, initiated by the Commission's Regional Policy Directorate (DGXVI) and closely connected with the emerging agenda of the Committee of the Regions, such attempts have begun. The UK's official position in relation to these initiatives is, not unusually, somewhat ambiguous. British pride in its long-standing land-use planning system suggests that it is 'bound' to be superior to those operating in other member states, and that any process of convergence must mean a reduction in British planning standards. Thus there are fears that as much might be lost as gained from a process of convergence. But the forthcoming 'Compendium of Planning Systems' will provide a better information base about the systems that currently apply, and will reveal differences in style between member states (formal *v* informal, flexible *v* prescriptive) as well as substance. British pride in its planning system may have blinded decision makers to the fact that other countries, while having a far less comprehensive system, are more prescriptive and can deliver objectives more reliably than the heavily discretionary UK system. This issue is likely to be a recurring feature of the inevitable debate about the merits or otherwise of greater cohesion in the land-use planning systems across Europe.

Environmental Impact Assessment

One reason for this explains the origin of the second bundle of issues, in which the European-originated Environmental Impact Assessment (EIA) system is prominent. The advent of the Single Market and its stimulus to transnational investment meant that member states with low environmental safeguards (for example, as operated by the land-use planning system) risked becoming dumping grounds for un-neighbourly activity such as polluting factories, waste disposal or other damaging developments. The 1985 EIA Directive (85/337) was therefore triggered in part by a desire to provide a consistent process for dealing with major developments that might have an adverse environmental impact. For practical reasons, the mechanisms introduced have mostly been grafted onto pre-existing land-use planning consent systems in individual member states, thus shaping and – to a degree – introducing one element of convergence in these systems. The current Directive only applies to projects, but a draft Directive on Strategic Environmental Impact Assessment, applying to plans, policies and programmes, agreed by the Commission in early 1997, would further challenge conventional land-use planning mechanisms.

The emergence of EU legislation on Environmental Impact Assessment provides an example of European levers being used to influence policy at the domestic level, and *vice versa*. During the late 1970s and early 1980s it became clear to UK environmental groups, who had long been arguing for better processes of prior environmental assessment of major developments, that simply by applying domestic pressure they were getting nowhere. Instead, they turned their attention to European policy makers, and were successful in persuading them that a European instrument, applicable to all member states, would bring real environmental benefits. Thus a draft directive on Environmental Impact Assessment was brought forward and finally approved in 1985. During negotiations, the UK Government had only reluctantly supported it, partly because it was convinced that existing domestic legislation was adequate to implement it without major reforms. In particular, the UK Government was convinced that Annex 2 of the Directive (covering projects where EIA was intended to be discretionary) would not need to be implemented in the UK unless the Government chose to do so. However, during a late stage in the negotiations, it became clear that special implementing instruments would be needed in relation to the (mandatory) Annex 1 projects; and, more seriously, that EIA for Annex 2 projects would not be discretionary for projects having a 'significant impact on the environment'. The groups, whose questions and probing had revealed the necessity of these more onerous requirements, were delighted that the Directive's practical effect would be wider than the UK Government had assumed, and pressed hard for implementation and enforcement to the letter.

There have been more complaints to the European Commission about the failure to implement the EIA Directive than any other piece of European legislation. This is partly because EIA is a new process, providing many 'hooks' on which compliance can be judged, and partly because its timing and importance opened many campaigners' eyes generally to the opportunities presented by lobbying in Europe. Many of the *causes célèbres* of the 1980s and 1990s, including the M3 extension through Twyford Down and the Newbury bypass, were the subject of complaints to the European Commission over alleged failures in the EIA process (see Chapter 1; Plate 3.1).

Small wonder, then, that attempts by the European Commission, widely supported by environmental groups, to extend the EIA legislation beyond projects to plans, policies and programmes, has led to resistance by member states, not least the UK. It does not want another stick with which environmental groups can beat it. However, its antipathy to further legislation in this field sits ill with the claim to be an environmentally responsible member state, so the Government has had to moderate its objections and a draft Directive now awaits scrutiny by the Council of Ministers.

EU funding mechanisms

Finally, in this area, huge controversy has accompanied some of the EU's own activities, especially where it has funded major development projects under the Structural and other Funds. In many cases these have been of a scale unlikely to have been feasible without EU assistance, in locations where their environmental implications raise serious problems, and have often been − or been felt to be − subject to seriously deficient consent processes. Examples include development along Mediterranean coasts, major road, bridge or rail infrastructure projects, tourist and leisure developments, and support for major economic projects and investment programmes, such as in Objective 1 and 5b areas. Amendments to Structural Fund procedures, following intense pressure from environmental groups, to ensure projects were subjected to more adequate environmental scrutiny, have not dispelled these anxieties (Long 1995).

FIONA REYNOLDS

As a result, attention is now being focused on the wider context within which such projects need to be viewed – what has come to be termed the 'spatial development perspective' – the belief that a better understanding of and clearer goals for spatial development will help to guide EU funding mechanisms and deliver better, less controversial but also more useful end results (see p. 142–3).

CONCLUSION

It has become clear that through its major interventions in some areas – agricultural and structural funding, for example – the European Union has a huge direct impact on land-use policy, though often by default and with significant disbenefits socially and environmentally. As awareness of these disbenefits grows, the emergence of a desire for a more positive and conscious land-use policy is not surprising, though its evolution is unlikely to be a smooth and predictable process. So just as member states are discovering the importance and interconnectedness of how people live, work, recreate and move around – and thus the vital role of land-use planning as the arbiter of many decisions about location and quality of life – so the engagement of the Commission and other EU institutions in this debate is not surprising.

Land use and landscape, though rarely articulated in precisely those terms, are at the heart of the debate about the future of environmental policy in Europe. Britain's long experience in these policy areas should be an asset. However, the UK Government has not been their champion in Europe, preferring to deal with issues such as CAP reform in narrowly economic terms, and to maintain national discretion for policy development and implementation in these other areas as far as possible.

This approach is unlikely to remain adequate or feasible as connections between different policy areas are made more and more explicit, not least by other actors on the scene – whether the environmental lobby or more recent arrivals such as local authorities through new mechanisms such as the Committee of the Regions (see Chapter 8). Recent and emerging developments, including those described above, mean there is likely to be a more structured and focused debate at EU level on land use and landscape over the coming years.

These issues are interesting precisely because they reach so widely into other policy areas and touch on some of the most fundamental questions which will have to be addressed by the EU. These include: What is it to be European? Do we have a shared cultural inheritance? What binds us together as European citizens and what common principles should we accept? Is land-use planning the essential missing ingredient in a coherent environmental policy? Thus landscape is a metaphor for what we value in our wider surroundings. Land-use planning mechanisms, in their broadest sense as a means by which we can articulate and better manage society's environmental responsibilities, may therefore have a powerful role to play in wider discussions about what tomorrow's Europe may mean for its citizens.

NOTES

1 The negotiating group for a group of smaller countries, such as Australia and New Zealand, that are heavily dependent on food exports and that are either dispensing with or have already removed state subsidies for agriculture.

14

WATER QUALITY

Neil Ward

Britain's relationship with European water quality policy has been one of the most turbulent and controversial aspects of the Europeanisation of environmental policy. Indeed, water pollution control is often used as a paradigm of the conflict between two differing 'styles' of environmental regulation. The British style is one characterised by flexibility, pragmatism, administrative discretion and an avoidance of absolute, statutory standards. This British tradition has been subject to increasing challenge and modification from an emerging 'European' style of environmental regulation based on formal and more legalistic systems of pollution control, with legal standards and timetables laid down for compliance with them.

The increasing importance since the mid-1980s of Community laws in protecting and improving water quality in the UK has also had important implications for the new, post-privatisation structure of water quality regulation in Britain and has helped contribute to a significant politicisation of water quality issues, particularly with respect to drinking water quality and waste water discharges to coastal waters. The UK has twice been found guilty of being in breach of EU water quality laws at the European Court of Justice and various environmental pressure groups have been able to exploit the new political spaces and opportunities opened up by the application for the first time of numerical water quality standards in the UK (Ward *et al*. 1995; 1996; Ward 1996b).

This chapter examines the process of European integration for the UK in the sphere of water quality policy. It describes the British tradition of water pollution control 'pre-Europeanisation' and outlines the milestones in the UK's experience with European water quality policy. The chapter then goes on to examine the impact of Europeanisation on British practice in terms of the substance of policy, the concepts and style of policy and the various organisational relationships in the water sector. Finally, recent developments in European water quality policy in the period since 1992 are assessed and the possible future implications for the UK are explored.

THE UK TRADITION IN WATER POLLUTION CONTROL

Britain's administrative system and style of environmental policy is seen to have a distinctive set of characteristics (see Vogel 1986; Lowe and Flynn 1989; Chapter 1). The first is a legislative system which has evolved from an accumulation of common laws, statutes, procedures and policies. Second is a regulatory system pervaded by administrative (and hence informal, accommodative and technocratic) procedures rather than judicial (and hence formal, confrontational and legalistic) procedures. Third, the British system has involved an avoidance of legislatively prescribed standards and quality objectives. Fourth, a voluntarist approach in dealing with private concerns has sought to foster co-operation and striven to achieve policy objectives through negotiation, persuasion and self-regulation. Finally, pollution control policy making has traditionally taken place in relatively closed policy communities, with policies usually agreed between civil servants and representatives of producer interests in private arenas largely invisible not just to the public, but also to Parliament.

In the sphere of *water quality management*, some aspects of what is conventionally understood as the 'British approach' to environmental policy appear stronger than others. Closed policy communities, a reluctance to impose fixed, statutory standards and a close relationship between the regulators and the regulated can all be clearly identified. But the water sector is distinctive from other areas of environmental policy, which have tended to develop as piecemeal and incremental responses to the emergence of specific environmental problems. In contrast, water management in the UK has long been organised around a very clear, logical and strategic framework informed by the principle of integrated river basin management (see Liddell 1974). The River Boards Act 1948 represented the first attempt at comprehensive river basin management and the approach was further enhanced when the Water Resources Act 1963 rationalised 34 river boards into 29 river authorities. By 1974 water resources in England and Wales were being managed under an even more rationalised and integrated system dominated by just 10 large regional water authorities (RWAs). The whole *raison d'être* of these reforms was the integration of water management based on the hydrological cycle. The RWAs were given responsibilities for all aspects of water management, including drinking water supply, sewage and waste water, and the regulation of pollution.

The political and economic context within which these regulatory characteristics have evolved is also worthy of comment. In the early 1970s, water pollution control was overhauled by the passing of the 1974 Control of Pollution Act and the establishment of the RWAs. Under the post-1974 regime, pressures from central government to control public spending posed tight constraints upon the investment priorities of the RWAs, and these constraints became particularly acute during the economic crisis of the late 1970s under the Labour Government and during the first Thatcher administration. As a result, although the British sewerage system – much of which was constructed in the Victorian period – had been subject to decades of under-investment (Kinnersley 1994; Maloney and Richardson 1995), successive governments in the 1970s and early 1980s delayed the full implementation of Part II of the Control of Pollution Act (dealing with the control of water pollution). The timescale for introducing legal powers under the Act was extended as pollution control policy became subordinated to public expenditure policy and a set of fears about overburdening industry with financially onerous regulatory controls in a time of recession (Levitt 1980). The discretionary nature of domestic legislation thus gave the government considerable leeway over the pace of its implementation. As the Department of Environment explained to the Royal Commission on Environmental Pollution in 1984:

> the plan for bringing COPA II into effect gives the government a considerable control over the rate of implementation on the ground. While Ministers will undoubtedly take steps to bring within the controls any type of discharge ... which may be defined by European Community directives as requiring to be controlled, they can be expected to have resource and economic considerations as well as environmental considerations very firmly in mind in deciding what more to bring under control, and ... at what time.
> (Royal Commission on Environmental Pollution 1984: 79)

This ability of the UK Government to 'take the foot off the accelerator' when it comes to environmental improvements has been progressively diminished by the growth, in both volume and impact, of European environmental legislation. Indeed, it was only in response to pressures generated by the need to comply with European laws that Part II of the Control of Pollution Act was eventually implemented in 1986 (Haigh 1989).

Coupled with this reluctance to incur additional spending on improving pollution control in order to meet the objectives of the European water quality Directives was a general complacency and underestimation of the extent to which European Directives were likely to impinge upon British practices and procedures. This complacency only began to be undermined in the mid- to late 1980s. Haigh and Lanigan have suggested that:

> the idea that British water might not be clean enough to pass tests which would also have to be met by continentals with supposedly dirtier water probably did not occur to the British government. In any case, the ... period allowed for implementation allowed time for any necessary adjustment, and there was always the probability that

delays and derogations (i.e. exemptions) would be granted in difficult cases.

(1995: 22)

In British water quality policy, elements of the characteristics outlined above currently remain, but by the late 1980s pressures had built up to stimulate a major reappraisal of policy and structures. In particular, there was a marked politicisation of water quality issues. The role of EU environmental policy has been, as we shall see below, an important stimulus to this politicisation, and has meant that British traditions have come increasingly to be challenged (Osborn 1992).

However, Europeanisation of water quality policy in the UK during the 1980s cannot be considered as a single, clearly discernible, process operating in isolation from other political and regulatory developments. Also of great importance was the decision by the Thatcher administration to transfer the ownership of the water supply and waste water treatment functions of the former RWAs from the public to the private sector. The move was arguably the most politically controversial of the Thatcher privatisations, and opened up the issue of water pollution and the UK's environmental regulatory record in this respect to unprecedented public and political scrutiny. It is, therefore, the *interplay* between Europeanisation and privatisation that has been crucial in shaping developments in water quality policy since the mid-1980s (see Table 14.1).

EUROPEAN WATER QUALITY POLICY

European water quality policy has been dominated by the Directive – a legally binding instrument in terms of the results to be achieved, but one which leaves to the member states the choice of methods enacted to ensure compliance. The first EC water Directives were adopted by the member states in the period 1975–6, coinciding with the European Commission's First Action Programme on the Environment. Discussions about the first major water quality Directive – the Surface Water for Drinking Directive (75/440) – began before the UK joined the Community on 1 January 1973, but the final draft of the Directive was not finally agreed until June 1975.

The Surface Water for Drinking Directive was designed to ensure that surface water abstracted for drinking met certain standards and was treated adequately before being supplied for consumption. Surface water sources were required to be classified into three categories – A1, A2 and A3 – according to their existing quality, with different levels of treatment required, and abstraction was prohibited from any waters whose quality was worse than A3. Lists of surface waters were sent by the UK Government to the Commission in 1981 and none was found to be worse than A3. Indeed,

Table 14.1 The UK and Europe: key developments in water quality policy since 1973

Year	UK	EU
1973	Water Act: Establishment of 10 Regional Water Authorities in England and Wales	
1974	Control of Pollution Act	
1975		Quality of Surface Waters for Abstraction Directive
1976		Bathing Water and Dangerous Substances Directives
1980		Drinking Water Directive
1986	First proposals to privatise the Regional Water Authorities published	
1988	Friends of the Earth study of pesticides in drinking water published	
1989	Water Act: Regional Water Authorities privatised and National Rivers Authority created	
1990	Environmental Protection Act	
1991		Urban Waste Water Treatment Directive
1992	Royal Commission on Environmental Pollution Sixteenth Report	Edinburgh Summit decision to review EU water policy
1994		New draft Bathing Water Directive published
1996	NRA and HMIP merge into Environment Agency	Review of EU water policy published

the negligible impact of the Directive in the UK was illustrated in the House of Commons in 1979 by a Minister who said that application of the Directive 'has caused no concern or difficulty, since the standards of water that are abstracted for drinking are at least as good and often better than those laid down in the Directive' (quoted in Haigh 1990: 38).

The next water Directive to be proposed by the Commission proved to be far more problematic for the UK and its relationship with the emerging European style of water quality regulation. Proposed in October 1974, and finally agreed by the member states in May 1976, the Dangerous Substances in Water Directive (76/464) brought to light a difference of view between the UK and the rest of the Community on the philosophy underpinning water pollution control (see Haigh 1990: chapter 3; Taylor *et al.* 1986). The Directive set a framework for the reduction of pollution of inland and coastal waters by particularly dangerous substances. An Annex to the Directive contained two lists of groups of dangerous substances. List I included organophosphorous compounds, carcinogens and mercury and cadmium compounds, while List II contained less dangerous compounds such as those of zinc, copper and lead. Discharges of substances on either list were to be subject to prior authorisation by a competent authority. Although Haigh (1990: 13) has warned that the dispute between the UK on the one hand and all the other member states and the Commission on the other has become caricatured and over-simplified by many commentators, it is true to say that during negotiations over the disputed part of the Directive dealing with List I substances, the UK was alone in preferring a regulatory approach based on setting individual emission standards by reference to local environmental quality. The rest of Europe preferred uniform emission standards as the principal tool for pollution control. In comparative European terms, Britain's short, fast-flowing rivers and turbulent tidal seas meant that uniform emission standards were considered to risk 'over-penalising' those polluters located on estuaries, where geography favoured the UK's traditional 'dilute and disperse' approach to pollution. Following a protracted and controversial disagreement, a compromise was accommodated in the Directive whereby two alternative regimes were included. A 'preferred regime' entails limit values for List I substances which emission standards are not to exceed, while an 'alternative regime' allows emission standards to be set by reference to environmental quality objectives. When the Directive was finally adopted, all member states except the UK declared that they would adopt the preferred regime.

In parallel with the negotiations over the Dangerous Substances in Water Directive were those around the Bathing Water Directive (76/160), proposed in February 1975 and agreed in December of that year. In the UK, this Directive has become one of the most widely known of Europe's environmental policies. It aims to protect the environment and public health by raising or maintaining the quality of bathing water over time. Waters used for public bathing must meet basic microbiological standards. If a bathing water falls below these standards, it must either be improved (usually by ensuring that sewage is not present or is adequately diluted) or closed to the public. Minimum sampling frequencies are laid down and member states are required to report to the Commission on bathing water quality.

The UK was not an enthusiastic supporter of the proposed Directive, and at the time the proposal was being discussed was, as a new member state, still struggling to come to terms with the policy process in Brussels. Directives tended to be seen by British officials as broad statements of intent, rather than binding pieces of legislation. Furthermore, the Department of the Environment (DoE) had given a firm defence of the use of environmental quality standards (as opposed to uniform emission limits) during the negotiation of the Dangerous Substances Directive, so found it difficult to reject such standards for bathing water. The UK therefore agreed

DANGER

DO NOT LOITER - SEWAGE HAZARD!

This beach is contaminated with sewage
from a nearby sewage outfall
which has caused high concentrations
of bacteria to be present.

DO NOT

ALLOW CHILDREN TO PLAY HERE · BATHE FROM
THIS BEACH · TOUCH OR PICK UP ITEMS

Plate 14.1 The full implementation and the strengthening of the Bathing Water
Directive have been taken up in campaigns by Greenpeace, Friends of the
Earth, Surfers Against Sewage, the Marine Conservation Society and the
Green Party.

Source: Greenpeace/Greig

to the adoption of the Directive in the Council of Ministers although, with hindsight, it can be seen that the approach embodied in the Directive would come to pose extensive problems for British practice.

Prior to the introduction of the Directive, sewage discharges to coastal waters were influenced not by any concern to protect the quality of bathing waters, but by a concern for cost effective sewage disposal, which often meant coastal discharge with no prior treatment. Advice published in 1975 by the DoE from the Coastal Pollution Research Committee had stated that 'The sea provides cost free and efficient sewage purification, and so long as this does not conflict with amenity or public health it seems sensible to take advantage of the fact' (DoE 1975: 3). Because of its faith in the 'dilute and disperse' approach, the UK was initially 'very critical' (Levitt 1980: 110) when the draft Bathing Water Directive was first proposed by the Commission. Indeed, the subsequent debate by the European Parliament in May 1975 consisted entirely of consideration of amendments proposed by UK delegates who stressed the practical and logistical difficulties involved in implementing the Directive as then drafted (see Levitt 1980: 95–113).

British concerns at that time focused upon the draft directive's 'practical implementability' (the definition of what counts as a 'bathing water', the methods of measurement and monitoring proposed, and the (initial) proposal that bathing be banned in those waters breaching the standards laid down in the Directive). At the same time, there was little realisation of the extent to which bathing waters in the UK might fail to comply with the Directive. Indeed, the then Environment Minister, when asked about the creation of league tables of beaches, replied that 'the truth of the matter is that until the [1974] Control of Pollution Act becomes operative and regional water authorities get operative we really do not know the extent of the problem in this country' (quoted in Levitt 1980: 98). By the mid-1990s, however, the expenditure required to ensure compliance with the Directive was running into billions of pounds.

The next Directive to have a substantial impact on UK regulatory practice was the Drinking Water Directive (80/778), proposed in 1975 and, after lengthy negotiations, finally agreed in 1980. The Directive was intended to standardise drinking water quality norms across the member states of the European Union in order to protect human health. Sixty-two different standards (or 'parameters') for water quality were laid down along with guidelines for water quality monitoring. Member states were permitted to apply more stringent provisions than those in the Directive if they so wished.

The Directive has provoked controversy in the UK (as well as elsewhere in Europe) because it has revealed the spread and levels of contamination of drinking water supplies with various contaminants, including nitrates and pesticides. Prior to implementation of the Directive, British legislation required that drinking water be 'wholesome', but what constituted 'wholesome' was not defined. Now 'wholesome' is defined in the statutory regulations laid down to transpose the requirements of the Drinking Water Directive into UK law.

The Directive has, therefore, become an important legislative component of the new, post-privatisation regulatory regime governing drinking water quality and the water companies have put in place investment schemes at water treatment works and in the distribution system to improve compliance with the various standards in the Directive. The improvements have required investment of the order of £2 billion since 1989 (DoE/Welsh Office 1993: 12), a considerable proportion of which has gone on installing equipment at water treatment works to remove pesticides from drinking water.

Probably the most important directive affecting water quality management to be passed since privatisation of the water industry in 1989 has been the Urban Waste Water Treatment Directive (91/271) adopted by the member states in May 1991. The Directive sets priorities for the treatment of sewage according to the size of the discharge and the type and sensitivity of the receiving waters, and provides that all significant discharges should be treated. It sets secondary treatment as the norm but provides for less stringent – but at least primary – treatment in areas with high natural dispersion, described in the Directive as 'Less Sensitive Areas'.[1] In areas designated as 'Sensitive Areas' more stringent (than secondary) sewage treatment will be required. The broad criteria for defining 'Sensitive' and 'Less Sensitive' areas are laid out in the Directive, but precise definitions are left to individual member states. To interpret the requirements of the Directive, the DoE established an Implementation Group of interested parties in 1991, two months before the final draft of the Directive was published. The Group consisted of representatives from the DoE, the National Rivers Authority, the Ministry of Agriculture, Fisheries and Food, the Scottish Office, the (Scottish) River Purification Boards, DoE Northern Ireland, OFWAT, the Water Services Association – which represents the privatised water companies – and the Treasury.

In September 1992 the National Rivers Authority recommended sixty-three areas be designated as Sensitive Areas under the Directive (and so requiring special additional sewage treatment beyond secondary level prior to discharge). Following negotiations within the Implementation Group, the DoE decided to identify only thirty-three waters for designation (on the basis of their eutrophic status). In these waters, the installation of phosphorous removal plants at sewage treatment works will be required. Fifty-eight Less Sensitive Areas (with high natural dispersion), where only primary sewage treatment will be required, were also identified.

The saga surrounding the designation of different zones under the Urban Waste Water Treatment Directive prompted some public and political controversy, not least because of the government's attempts, eventually thwarted in 1996 by a judicial review, to classify parts of the Humber and Severn estuaries as high natural dispersion areas in order to avoid the requirement that sewage treatment be upgraded. This practice of seeking to minimise the financial impacts of Directives through tightly restricting the coverage of areas designated for special protection (and, in the case of the Urban Waste Water Treatment Directive, of maximising the coverage of those zones exempted from special protection) is far from new. For example, in implementing the Bathing Water Directive, the UK Government aroused controversy for its seeming circumvention of the Directive.

The Bathing Water Directive defined bathing waters as 'all running or still fresh waters or parts thereof and sea water, in which: bathing is explicitly authorised by the competent authorities of each member state; or bathing is not prohibited and is traditionally practised by large numbers of bathers'. Because there were no statutory provisions by which public bodies either explicitly authorised or prohibited bathing, the government in 1979 asked the RWAs and district councils to identify those waters where bathing was 'traditionally practised by a large number of bathers'. A government Advice Note set out the criteria upon which waters were to be designated. These were: bathing waters with fewer than 500 people in the water at any one time would not be designated; stretches of bathing waters would be designated where the number of bathers exceeded 1,500 per mile; and bathing waters with between 750 and 1,500 bathers per mile would be open to negotiation between the RWAs and district councils (NRA 1991: 24). This very narrow interpretation of what constitutes a 'bathing water' resulted in the designation of only twenty-seven waters for the whole of the UK, excluded many well-known bathing resorts and included no inland bathing sites. This initial list, which meant that the UK had fewer designated bathing waters than landlocked Luxembourg, drew derision from influential commentators and the government's independent advisors (Royal Commission on Environmental Pollution 1984; Haigh 1990). The European Commission issued a Reasoned Opinion that the UK had failed to take all the necessary steps to comply with the Directive (OJ, C8/21, 10.1.85) and threatened infringement procedures. This move, combined with the public and political outcry over the designations, forced the government to widen its definition. Less emphasis was placed on strict criteria relating to the number of bathers, and instead factors such as the presence of lifeguards and the provision of changing huts, car parks and toilets were used to identify bathing waters for designation. As a result, the Government announced in 1987 that an additional 362 bathing waters had been identified for the purpose of the Directive.

A similar strategy was pursued with the Shellfish Waters Directive (79/923), agreed by the member states in 1979 and designed to maintain or improve the quality of shellfish waters. Here the government's thinking was explicitly stated in a DoE Advice Note issued to the RWAs in 1980, which instructed them thus:

In the current economic circumstances the Government accepts that the implementation of the Directive should not have an undue effect on water authorities' capital expenditure plans. For this reason water authorities should aim, in the initial round to designate only a small number of waters which either already meet the appropriate standards or which are capable of doing so by October 1987 after improvements which are already programmed. [*sic*]

(DoE 1980: 1–2, quoted in Ward *et al*. 1995: 12)

As a result of this approach to designation, only twenty-seven shellfish waters were initially designated under the Directive, representing only a small proportion of the 500 or so sites around the UK where shellfish are commercially grown.

Despite such strategies to limit the impact of water quality Directives through the process of designation, it is fair to say that British water quality management has been substantially overhauled as a result of the requirements of European Directives. The 1991 Urban Waste Water Treatment Directive can be argued to represent the 'high point' of European water quality policy in terms of its impact on British water pollution control policy and practice. It has meant, together with the requirements of the Bathing Water Directive, that the amount of money currently being spent on improving sewage discharges in the UK to improve compliance is staggering – £9.5 billion between 1989 and 2005 (House of Lords Select Committee on the European Communities 1995: 6). We return later to consider events since this 'high point'.

THE IMPACT OF EUROPEANISATION ON UK PRACTICE

The substance of policy

As was suggested above, it is at times difficult to disentangle the impacts of European Directives *per se* from the implications of water privatisation. Despite this difficulty, we might point to the following, profound implications of 'Europeanisation' for the *substance* of water quality policy in Britain.

First, European Directives have required that absolute legal standards be put in place for a range of water quality parameters. Most important have been the sixty-two parameters laid down in the Drinking Water Directive, the bacteriological parameters laid down in the Bathing Water Directive and those parameters relating to the discharge of dangerous substances to water. There is little doubt, as far as the water sector is concerned, that these environmental standards are stricter in the UK than would have been the case in the absence of European legislation and that their existence has forced the pace of investment in environmental improvements in the water industry.

Second, the requirement that water be monitored (whether, for example, this be drinking water supplied at customers' taps, or bathing waters monitored at predefined points) and that monitoring results be assessed to measure compliance against these standards has brought new types of water quality problems to light. For example, it is only since implementation of the Drinking Water Directive that the spread and levels of contamination of drinking water supplies by pesticides and nitrates have been recognised, prompting new policies to remove these contaminants from supplies and increased public and political pressure for greater controls over their use (Ward et al. 1995). The scale of nitrate contamination revealed by the monitoring under the Drinking Water Directive has helped prompt the development and adoption of a new directive – the Nitrates Directive (91/676) – specifically designed to address the sources of nitrate pollution across Europe. Similarly, the extent of pesticide contamination revealed by Drinking Water Directive monitoring has prompted pressures for greater controls over pesticide use. Friends of the Earth have collated the monitoring data collected to meet the requirements of the Drinking Water Directive and have revealed that, in 1992, 14.5 million people in England and Wales lived in areas supplied with drinking water in which pesticides were, at times, present at levels higher than those stipulated in the Directive. As a result of this recognition of the scale of contamination, some pesticides have now been withdrawn from use in the UK, the use of others has come under renewed review, and greater attention has been given to policies and programmes which encourage the minimisation of pesticide use and the adoption of integrated pest management practices in agriculture (Ward 1996b). The monitoring required under EU water Directives has, therefore, begun in recent years gradually to shift attention in the UK from the regulation of point of source pollution to less visible, longer-term and chronic forms of diffuse pollution, particularly those associated with agricultural land uses.

Third, it is inconceivable that we in the UK would have had such an extensive clean-up programme to improve the quality of coastal sewage discharges without the pressures brought about as a result of implementation of the Bathing Water Directive. By the mid-1990s, current and planned expenditure ran to almost £10 billion, although these monies were for sewage treatment schemes that help meet the requirements of both the Bathing Water Directive and the Urban Waste Water Treatment Directive. Nigel Haigh (1994a: 4.5–7) has highlighted how the Bathing Water Directive has 'led to the commitment of *substantially greater expenditure on sewage treatment and disposal than would otherwise have occurred*' (emphasis added), with the result that in England and Wales the level of compliance with the mandatory parameters in the Directive has risen from 62 per cent in 1987 to 89 per cent in 1995.

Finally, and perhaps most important, the Europeanisation of water quality management has been an important influence upon not only the decision to privatise the water industry in England and Wales, but also the eventual structure of the privatised industry and its various regulatory relationships. Maloney and Richardson, in their detailed study of policy change in the water industry, have suggested that the prospective investment programmes required by EU directives were 'probably the most important factor in the Government's decision to privatise water' (1995: 58). By 1989, the decades of neglect in upgrading water infrastructure were such that the DoE estimated a ten-year investment programme of £24.6 billion would be required. Passing on these obligations to newly privatised water companies therefore had obvious appeal.

Moreover, it was largely as a result of the influence of European law that we saw, in 1989, the formation of the National Rivers Authority which, significantly, styled itself as 'Europe's strongest environmental protection agency'.[2] The original proposals for privatisation of the water industry were for the RWAs to be privatised *in toto* (including their pollution control regulatory functions). Environmental organisations such as the Council for the Protection of Rural England and the Institute for European Environmental Policy questioned whether private water companies could constitute competent authorities to implement European legislation. Following a 'carefully worded warning' (Maloney and Richardson 1994: 123) from the European Commission on this issue, the original privatisation proposals were shelved and subsequently revised to include the creation of the National Rivers Authority.

The concepts and style of policy

Europeanisation has led to the gradual adoption of new concepts in environmental policy in the UK and has helped foster important shifts in the style of policy – the characteristic operating procedures adopted by the relevant organisations. Probably of greatest significance is the shift from flexibility to formality in establishing the objectives of water pollution control policy. Thus where polluting discharges were previously set in the context of local environmental conditions and the uses made of water courses, now standards are set to accommodate the statutory requirements of European legislation. The introduction of such standards where once there were none represents a radical departure for British water pollution regulation.

Prior to the development of a significant water quality policy for Europe, the British approach to water pollution control, particularly the long-unimplemented Part II of the Control of Pollution Act 1974, tended to allow central government to subordinate environmental improvements to the priorities of public expenditure policy. A common European framework for water quality management, coupled with legal standards to be met over timescales that were laid down in legislation, has greatly reduced the UK Government's freedom of movement in this respect. In addition, the justification for many of the water quality standards in European legislation has drawn upon traditions and principles of environmental management that differ markedly from the British approach.

In particular, European regulatory philosophy draws upon the principle of precautionary action – a principle whose origins lie in the 1970s *Vorsorgeprinzip* (the principle of foresight) in German pollution control. The precautionary principle has evolved to become an increasingly central goal of Community environmental policy, particularly since the Fifth Environmental Action Programme, but jars with the continued British enthusiasm for 'sound science' to 'prove' causes and effects in advance of regulatory action. Enthusiasm for the principle among European institutions and other member states has made it increasingly difficult (although has not prevented it completely) for the UK Government to call for the repeal of some standards (such as those for pesticides in drinking water, for example) in the absence of thorough scientific and toxicological justification in terms of threats to public health. This and other examples – such as the repeated questioning of the scientific basis for the mandatory coliform standards in the Bathing Water Directive – suggest that the UK's commitment to the precautionary principle has been more rhetorical than substantive.

Finally, it is worth noting that there is a risk of overplaying the presumed 'loss of discretion' implied by the Europeanisation of water quality policy in the UK. Elements of Britain's traditional pragmatism and 'flexibility' persist through the detailed process of implementation (Ward *et al.* 1995). For example, as we saw above, discretion has still been exercised through the process by which different geographical zones are designated for particular forms of protection under various Directives. Examples here include the Bathing Water Directive, the Shellfish Waters Directive, the Nitrates Directive, and the Urban Waste Water Treatment Directive. In each of these cases, the areas designated for protection have been tightly constrained – sometimes (as in the case of the initial designations under the Bathing Water Directive) to the point of absurdity. In each of the examples listed above, a strictly minimalist approach in the UK to designating zones has diminished the scope of environmental protection the Directives could, potentially, have offered.

Organisational relationships

Europeanisation has helped open up and destabilise the water policy community and this has been crucially helped by the privatisation of the water industry in the UK (Maloney and Richardson 1994; 1995). A much wider range of actors are now included in the policy-making process compared to the closed policy community of technical specialists and regulatory officials

in the period prior to the 1980s. New participants include, of course, officials from the European Commission and representatives of EU-wide
industry associations, but also environmental and consumer groups. Thus,
for example, when the European Commission announced its intention to
review the Drinking Water Directive in 1993, a much wider range of actors
was consulted than had previously been the case under British, or even early
European, legislation.

In September 1993 the European Commission hosted a conference in
Brussels where over 260 participants from a range of interested parties met
to discuss the review of the Drinking Water Directive. Participants included
representatives from the Commission, the European Parliament, member
state governments and regulatory authorities, water suppliers, agricultural
and industry interests, environmental and consumer pressure groups and a
range of research institutes from across Europe. The eventual publication of a revised draft Drinking Water Directive in January 1995 suggested
that UK Government attempts to effect a significant change in the content of the Directive, especially regarding the controversial pesticides para-
meter, met with little success in the face of concerted opposition from
environmental groups and other member state governments (Matthews and
Pickering 1996).

The water policy community has also been opened up in the UK as a
result of the rise of the environmental pressure group sector and its increasing legal and technical expertise, as well as by a distinctive *politicisation* of
water quality issues during the 1980s (Lowe *et al.* 1997). European water
quality Directives, coupled with privatisation, have provided what political
scientist Hugh Ward (1993: 135) calls 'a unique political opportunity structure'. This has enabled environmental pressure groups in the UK to 'Euro-
peanise' their campaigns and point to breaches of directives as breaches of
European law. Absolute numerical water quality standards, combined with
the requirement that drinking water and bathing waters, for example, be
monitored, provide the statistical information and a campaigning 'yardstick'
with which environmental groups have been able with good effect to bring
pressure to bear on traditional British complacency regarding pollution from
nitrates and pesticides and from sewage discharges to sea. Thus, for
example, Friends of the Earth could point to the presence of pesticides at
levels higher than the Drinking Water Directive's standards across much of
England as a *widespread breach of European law*. The moral opprobrium in
Britain associated with such *illegality* has certainly been an important stimulus
for action to reduce the scale of contamination. Moreover, Europeanisation
has also helped open up the space for new forms of political activity at the
local level. For example, the combination of fixed legal standards for bathing water quality and the requirement that bathing waters be regularly monitored and that monitoring results be made publicly available has provided
the ammunition for locally based environmentalists to bring pressure to bear

upon water companies to improve sewage treatment. Examples of such campaigns include those of local Green Parties and of the environmental pressure group Surfers Against Sewage (SAS). In the southwest of England, a region plagued with a tradition of discharging raw sewage direct to sea, and containing over a third of the nation's designated bathing waters, local environmental activists have been successful in drawing national attention to 'dirty beaches' which *fail EC health standards* (see Ward *et al.* 1995; 1996). Similarly, SAS, who claim to be Britain's fastest-growing environmental pressure group, have attracted widespread media, public and political attention to the sewage and bathing water issue through their publicity stunts featuring surfers in gas masks accompanied by a large inflatable turd at 'failed' beaches.

Surprisingly, however, it is only very recently that British environmental pressure groups have specifically sought to influence European-level policy making with respect to water quality policy. Surfers Against Sewage, for example, following their formation in 1990, had been campaigning for three years before they eventually began lobbying European rather than British institutions. Indeed, it has primarily been through a whistle-blowing role, making use of European legal procedures for addressing noncompliance with EU laws, that British environmental groups have brought pressure to bear within the UK (see Chapter 5). In shaping the development of new EU water policies (or the review of existing legislation) environmental groups have struggled in their Brussels-based lobbying compared to the much better resourced representatives of polluting industries.

FUTURE DIRECTIONS

In this final section we turn to consider events since the Urban Waste Water Treatment Directive of 1991, which was the 'high point' of European water quality policy in terms of its impact on British policy and practice. Since this time, we can see a greater convergence of principles and approaches between the UK and the rest of Europe in the water policy sphere, and so less potential for conflict and controversy. This trend has, it should be noted, been attributed in part to changes in the personalities involved within the Commission (Haigh and Lanigan 1995). Of greater importance, however, are the wider shifts in the confidence in, and momentum behind, the process of European integration which lie beyond the environmental sector.

The political fallout from the troubled process of the ratification of the Maastricht Treaty has, since 1992, pervaded debates about the future of Europe's water quality policy. The initial Danish 'No' vote against Maastricht was taken at the time to imply popular concern about Europe's over-centralising tendencies and served to raise the profile of the principle of subsidiarity in European politics. Amid widespread fears among environmentalists that environmental policy would top the subsidiarity hit-list with the repeal of Directives and the repatriation of some standards from the EU level to the member states, the UK Government drew up a list of more than two dozen pieces of environmental legislation it wanted to see repealed or modified under the principle of subsidiarity. The list included several water quality directives. With similar demands coming from France, the Commission agreed to review its water Directives and to make amendments in the light of the subsidiarity doctrine, new scientific knowledge and technical advance.

The review has proved to be the most important development in European water quality policy since the first major directives were adopted in the 1970s and early 1980s. It has concentrated on simplifying and streamlining existing Directives but has opened up the whole rationale for Europe's water policy to greater public and political debate. Environmental groups have repeatedly expressed concern that the review poses a threat that standards will be relaxed, fears inflamed in part by claims by British ministers, in particular, that major concessions on the repeal of over-prescriptive European legislation had been won. However, overhauling European water policy has proved to be an intricate and protracted process. Whether fears of a weakened European policy are well founded or not will only become clear once the final drafts of new or reformed Directives have been agreed by the member states.

The review has been inching towards the withdrawal and replacement of some older Directives. At the European Council in Brussels in December 1993 it was announced that future water protection would be based on two sets of Directives – one dealing with water quality and including Directives on drinking water, surface water, bathing waters and groundwaters, and one with the control of discharges. Those Directives protecting surface waters for abstraction, freshwater fish and shellfish, for example, are to be replaced with a 'framework' Directive on the ecological quality of surface waters, a draft of which was published in August 1994. Such framework Directives are designed to give member states greater flexibility in deciding how they are implemented. Moreover, applying subsidiarity to the proposed Ecological Quality of Surface Waters Directive has meant the proposed legislation also gives member states greater discretion in setting targets for improving water quality as well as the timetable to achieve them. The fundamental conflict between subsidiarity and the notion of a level playing-field comes into sharp focus here. The original rationale for European-wide Directives was to harmonise environmental standards, while subsidiarity often seems to pull in the other direction. The Directive also appears to illustrate a convergence with the traditional British approach because of its greater emphasis on flexibility and national discretion and the fact that no absolute numerical standards are contained within it.

The reviews of the Drinking Water and Bathing Water Directives have concentrated on individual standards in the Directives. Much of the debate about the appropriateness of particular standards has been framed in purely technical terms, however, with lobbyists and policy makers arguing over what levels of contamination are 'scientifically proven' to be 'safe', particularly for pesticides in drinking water and sewage-related contaminants in bathing waters. As we have seen, complying with both Directives has entailed major costs which have meant rapidly rising water bills for consumers and raised the political temperature of water policy.

The water policy review is taking place under the new, post-Maastricht legislative arrangements and this has not only placed a greater emphasis on the role of the European Parliament, but has also had the effect of slowing the pace of reform. In February 1994, the Commission put forward proposals for a new Bathing Water Directive, dropping some scientifically less well regarded parameters and tightening others. The proposals were agreed by the Commission, the Council of Ministers and the Economic and Social Committee, but were rejected by the European Parliament's Environment Committee when it considered them in December 1994 (see NRA 1995: 23). The Committee felt that the scientific justification for the new standards and sampling methods was poor. As a result, the proposals have had to be returned to the Council of Ministers for reconsideration. It is likely to be some time before the political discussions between the member states leading to an agreement on the Directive's final draft are completed, meaning that the new Directive is unlikely to be in force until after the end of the century.

With the Drinking Water Directive, the Commission has emphasised that the review need not lead to a relaxation of standards, has reiterated the view that pesticides have no place in drinking water and has called the Directive's strict 0.1 microgramme per litre standard – originally a surrogate zero – an early example of the precautionary principle now embodied in the Maastricht Treaty. Industry interests and some member state governments have argued that the costs of complying with strict environmental standards can only be justified if comparable benefits are achieved. The UK Government, for example, complained of the 'over-stringency' of the pesticide standards, arguing that if the standards are too strict, expenditure on water treatment to meet them cannot be justified. Environmental groups, on the other hand, argue that it is impossible to derive suitable cost-benefit studies for the pesticide standard because there is no way of quantifying the benefits. When the proposed new Drinking Water Directive was circulated in draft form in January 1995, although the number of mandatory parameters had been reduced from 62 to 44, the strict standard for individual pesticides remained in place (House of Lords Select Committee on the European Communities 1996).

In effect, the prospect of European water quality standards being 'relaxed' appears politically to worry some member states and the Commission enough to ensure that subsidiarity should not result in a *general* weakening of established directives. Moreover, the inclusion of three new member states with strong environmental agendas and the greater role of the Parliament's Environment Committee ought to help ease environmentalists' recent fears that subsidiarity will be employed to unravel Europe's water policy. The Environment Committee hosted an expert hearing on water policy in June 1995 where, in contrast to calls for deregulation from industry, the Committee's chairman suggested the dominant problem remains the continued threat to water quality in Europe and the extent to which the EU is dependent upon the member states themselves to interpret, implement and monitor water legislation, implying that rationalisation of water policy should mean a clearer and stronger role for a common European framework. Thus, while 1992 and 1993 seemed to represent a window of opportunity for those concerned with the costs of meeting Europe's water quality standards to press for reducing their scope, events since then and the changing political complexion of the Union suggest that the window may be closing, making any substantive dilution of water policy an increasingly unlikely possibility.

Finally, there are other signs which suggest a convergence of views and priorities between the UK, European institutions and other member states, with the likelihood of a less acrimonious relationship between UK and European water policy in the coming years. The first of these is the greater efforts on the part of the Commission at *consultation* when drawing up draft legislative proposals. (Witness, for example, the different responses of the UK Government and industry interests to the revised Bathing Waters Directive (House of Lords Select Committee on the European Communities 1994) and the revised Drinking Water Directive (1996).) It had been a complaint in the past among British interests that draft water legislation appeared to 'come from nowhere', whereas now the Commission seems to be much more willing to outline its thinking and consult widely in advance of publication of legislative proposals. A second sign is the increasing concern across Europe for the practical implementability of legislation 'on the ground', as opposed to previous dominant concerns with the formal transposability of Directives into national legal systems. Steps to improve understanding of implementation difficulties and to share best practices have been made through the establishment of a network of implementing authorities in the environmental sphere, an idea claimed as a British initiative in British official circles (though claimed to be a French idea in French circles!). Finally, the increasing emphasis on the principle of subsidiarity in recent years has at least required a more explicit justification of regulatory action at the European level regarding water quality in accordance with British concerns.

In February 1996, the Commission issued a Communication which provided a wide-ranging review of EU water policy (CEC 1996e). The paper

Table 14.2 The principles of Community water policy

(i)	A high level of protection, rather than a minimum acceptable level, should be aimed for.
(ii)	The precautionary principle requires that where scientific knowledge is incomplete policy should err on the side of caution.
(iii)	Policy should recognise the moral duty to *prevent* damage to the environment occurring in the first place.
(iv)	Environmental damage should be rectified at source rather than seeking technical solutions to solve the problem 'downstream'.
(v)	The costs of measures to prevent pollution should be borne by the polluter.
(vi)	Water management objectives should be integrated into other policy sectors such as land use planning and agriculture.
(vii)	Policy should be informed by the best use of available scientific knowledge.
(viii)	Policy must be sufficiently flexible to avoid the imposition of inappropriate or unnecessarily strict requirements simply for the sake of harmonisation.
(ix)	The costs and benefits of action or inaction should be taken into account in determining water policy objectives.
(x)	Economic and social issues relating to regional and sustainable development should be considered in the development of water policy.
(xi)	Water policy should recognise the need for international co-operation.
(xii)	The principle of subsidiarity should be considered in the development of EU water policy.

Source: CEC (1996e)

outlined twelve principles seen to underlie Community water policy (see Table 14.2).

Of these principles, it is the seventh, eighth, ninth and twelfth that reflect sentiments that have been voiced particularly strongly in recent years by the UK Government and British industrial interests. Since water privatisation, the private water companies in England and Wales, through their umbrella body, the Water Services Association, have stepped up their efforts at lobbying at the EU policy-making level and, through their prominent role in two European-wide trade groupings – EUREAU and the European Waste Water Group – have sought to stress the financial implications of EU water proposals and to question the emphasis on the principle of harmonising standards (see Chapter 9). Aspects of the new approach to EU water policy would seem, therefore, to signal some degree of convergence between EU and UK agendas.

ACKNOWLEDGEMENTS

I would like to thank Philip Lowe and Steve Ward for their comments on a draft of this chapter.

NOTES

1 Sewage can be treated to different levels. Preliminary treatment involves the screening out of larger solids, and primary treatment usually includes a sedimentation process to allow sewage solids to settle out. Secondary treatment involves biological oxidation, using percolating filters or aeration in activated sludge tanks, followed by further sedimentation. A final stage, tertiary treatment, can be used to reduce the nutrient load in sewage effluent.
2 Lord Crickhowell, National Rivers Authority Chairman, quoted in the NRA's Summary Corporate plan for 1992/3.

15

INTEGRATING POLLUTION CONTROL

Jim Skea and Adrian Smith

Industrial pollution control in the UK has been modified considerably by European legislation. UK practice has long relied on the exercise of site-by-site administrative discretion in interpreting loosely defined legal principles. The conventional wisdom is that this traditional UK practice has been bending under the force of European legislation based on tightly specified emission limits which are inflexible and applied uniformly across industrial establishments (O'Riordan and Weale 1989; Jordan 1993). However, UK policy in this area has recently recovered a degree of self-assurance. A new framework European Union (EU) Directive on Integrated Pollution Prevention and Control (IPPC) (96/61) is said to draw heavily on UK experience with the domestic Integrated Pollution Control (IPC) regime. What does this signal about Britain's shifting relationship with the EU in the environmental policy domain? The aims of this chapter are to uncover the degree of UK influence on European legislation, to assess the extent to which Europe has influenced (and continues to influence) UK practice and to consider whether Britain's often troubled relationship with the EU has now entered a new phase with respect to this aspect of environmental policy.

Controlling the environmental impacts of industrial activity has a long tradition in the UK and other countries. British controls date back to the 1863 Alkali Act. France developed a law on classified installations in 1917. A common feature of various national systems has been the definition of general principles of control in primary legislation ('best practicable means' in the UK, 'state-of-the-art' in Germany) supplemented by legal and/or administrative mechanisms for developing and applying more specific controls at the sector and site level. The UK is not unique in having had difficulties in implementing European law. France, for example, had problems with the European Commission because it used administrative circulars to implement European law prior to 1991. At least some of the tensions apparent in the development of pollution control within the EU can be seen as operating along an EU–member state rather than a UK–Europe axis.

The new IPPC Directive reflects an attempt to harmonise procedures across Europe as well as two significant advances in the philosophy of pollution control. The first advance is the attempt to regulate releases to all environmental media – air, water and land – within a single framework. This makes it possible to take into account transfers of pollution from one environmental medium to another. The second feature is the promotion of preventive approaches to environmental management. This reflects a wide dissatisfaction with the less desirable aspects of traditional 'command-and-control' approaches to regulation. These are believed to have promoted the adoption of 'end-of-pipe' control techniques which can transfer releases from one environmental medium to another and inevitably add to production costs, thus eroding industrial competitiveness. Both policy makers and business are interested in developing new approaches which will promote the adoption of 'clean technologies' (CEC 1993b; Schmidheiny 1992). A move to preventive approaches is signalled in the UK Environment Act 1995, which led to the establishment of the Environment Agency in England and Wales and the Scottish Environmental Protection Agency (SEPA).[1]

The next section of this chapter describes the 'baseline' UK approach to pollution control, essentially the situation in the late 1970s/early 1980s. European influence through the 1980s, and the interplay of external pressures and domestic policy developments in the establishment of the UK IPC system are then discussed. The chapter considers the extent to which the formal introduction of IPC has affected the actual practice of pollution control in the UK. This leads into a description of the principal features of the new European IPPC system and the key debates that characterised its negotiation. Finally, the chapter addresses the impact of the UK approach on the new European regime; the extent of European influence on British practice, and the UK's changing relationship with Europe in this policy area. These are particularly important issues as responsibility for British pollution control is transferred to the Environment Agency and SEPA. A theme running throughout the chapter is the degree to which formal legal frameworks affect practice on the ground. Is policy made through Brussels Directives or does it emerge through site-level implementation?

POLLUTION CONTROL IN THE UK: THE TRADITIONAL APPROACH

Traditionally, the UK used single-medium regulations for controlling industrial releases. Separate regulators dealt with emissions to air, discharges to water, and disposal to land. This section focuses upon the air pollution regime which was typical of the UK regulatory style (Vogel 1986).

Industrial air pollution was traditionally controlled by the Alkali Inspectorate. The unit of regulation was the industrial process, and the regulatory duty was to operate processes according to a statutory principle rather than to statutory limits. Processes 'scheduled' for regulation had to be registered with the Inspectorate, which in turn had a duty to ensure that operators were employing the 'best practicable means' (BPM) for controlling emissions. The Inspectorate enjoyed considerable discretion over determining and enforcing BPM. They were considered the 'technical experts' by government and were allowed to operate quite autonomously. Within the Inspectorate, discretion was further devolved to field staff who were given the flexibility and authority to negotiate controls with operators, taking site-specific factors into account (Alkali and Clean Air Inspectorate 1974).

BPM was considered an 'elastic band' by the Inspectorate, providing them with the flexibility to tighten standards to reflect developments in pollution control (Ashby and Anderson 1981: 40). Standards 'presumed' to represent BPM for new processes were made public in BPM Notes for each type of scheduled process. However, existing processes were allowed to operate below these standards for the remainder of their economic lives (Tunnicliffe 1975). In other instances inspectors would relax an operator's timetable for improving emission controls if they had not been met, rather than shut the process down (Mahler 1967: 42). BPM Notes were essentially the public face of a closed, site-specific regulatory regime.

Inspectors were chemical engineers with at least five years industrial experience (Health and Safety Executive 1986: 2). The background shared by inspectors and operators meant that pollution control was regarded as a problem with purely technical dimensions (Weait 1989: 57). Regulator and regulated worked in partnership, tackling pollution problems through co-operation and reaching solutions through consensus (Ireland 1967). This was just as true for BPM Note production – where trade associations were invited to help draft standards (Alkali and Clean Air Inspectorate 1974: 12) – as it was for the site-specific negotiation of emission controls.

Inspectors sought levels of pollution control with which individual operators would comply voluntarily. Field experience meant that inspectors were familiar with good practice, and they would use this knowledge to encourage operators to move towards the best practice being achieved by others. This voluntaristic approach contrasts with a legalistic approach in which fixed emission limits are enforced.

There were practical reasons for the voluntaristic approach. First, the BPM principle was difficult to enforce. Emission limits in BPM Notes had no legal standing. BPM implied that a balance had to be struck between an operator's technical ability to control emissions and their financial ability to do so. BPM was site-specific, which meant that operators, not inspectors, were the source of the technical and financial information necessary to determine BPM. Thus, a second related reason for voluntarism was industry's possession of vital regulatory information, which assured them participation in negotiations over standards. Informal agreements over BPM were struck between inspectors and operators. The Inspectorate kept few documentary records concerning the BPM conditions for individual processes (National Audit Office 1991: 2).

Consequently, prosecutions were rare; because it was difficult to establish, in a legal sense, that BPM had been transgressed, because there was a risk to inspector credibility if a case was lost, and because prosecution would undermine an otherwise co-operative relationship (Weait 1989: 67–8). A final reason for seeking a voluntaristic approach was that the small Inspectorate was organisationally constrained. Even in the late 1970s, forty-three inspectors had to regulate 2,500 industrial processes. Inspectors could not monitor all processes at all times and public complaint could alert them only to visible or odorous emissions. Site inspections primarily provided an opportunity for inspectors to meet operators and build up a co-operative, problem-solving approach to pollution control.

To critics excluded from this air pollution regime the voluntary approach to compliance appeared like regulatory capture – a concern taken up by 'public interest' groups such as Social Audit (Frankel 1974). Inspectors were accused of helping 'industry to get on with the job' (Bugler 1972: 11), which 'meant the public interest in a healthy environment was not being sufficiently served' (Frankel 1974: 46). Critics also argued that inspectors were not tough enough on enforcement (Tinker 1972). Annual prosecutions rarely reached double figures – a sign of success to the Inspectorate but a sign of weakness to critics.

These criticisms prompted the Royal Commission on Environmental Pollution to investigate air pollution control (Ashby and Anderson 1981: 133). Its 1976 report recommended a more open and accountable approach to air pollution control. It argued that release limits should be legally binding and publicly available. It also recommended an 'integrated' approach, taking account of releases to all environmental media within a single regulatory framework (Royal Commission on Environmental Pollution 1976). This report was a landmark in UK thinking about integrated pollution control, though the cross-media environmental impacts were also being considered by the Organisation for Economic Co-operation and Development (OECD) and by the European Community (EC) under its Environmental Action Programmes. The Royal Commission report was rejected by the UK Government in 1982, but provided the intellectual arguments for the regulatory projects of the late 1980s.

EUROPEAN DEVELOPMENTS

The European decade 1982–93

Until the 1980s, the UK's pollution policy was an internal affair. Two factors began to draw the UK closer to its European neighbours. The first was the emergence of transboundary air pollution as an international environmental issue. Concern about acid rain led to British pollution becoming an issue of legitimate concern for other countries.

The second, and arguably more far-reaching, factor was European integration. In the field of pollution control, this was manifested in a push to 'level the playing-field' in relation to regulatory burdens experienced by industry operating in different parts of the Community. Germany was a particularly strong supporter of these developments because higher environmental standards, implemented through more formal legalistic means, were perceived to place German industry at a competitive disadvantage (Boehmer-Christiansen and Skea 1991). Britain came under considerable European pressure to renovate the ambition of its environmental goals and the means through which environmental policy was implemented (Haigh 1987). Specifically, there was pressure to replace administrative discretion with formal legal approaches.

The first challenge came with negotiations on the EC's 1984 Air Framework Directive (84/360). This was a single-medium measure intended to harmonise national approaches to the authorisation of major industrial sites. Germany had been keen to use the concept of *best available technology* (BAT), drawn from its domestic legislation, as the fundamental principle of the European Directive. The UK feared the technical prescription this implied and helped to formulate the *best available technology not entailing excessive costs* (Euro-BATNEEC) principle which was finally adopted. The Directive required member states 'to implement policies and strategies . . . for the gradual adaptation of existing plant . . . to the best available technology', implicitly challenging the more indulgent view taken by UK pollution inspectors. The Directive also challenged the UK discretionary approach by enabling the Council of Ministers to fix emission limit values.

Politically, the negotiations on the Air Framework Directive were a prologue to the five-year battle on the Large Combustion Plants Directive (88/609) which was agreed in 1988. This was a 'daughter' of the Air Framework Directive, containing emission limits for new power stations and other major sources. It also established national quotas for emissions from existing plants, placing obligations on the UK which it did not have the legal capacity to implement. There was considerable resistance to the Large Combustion Plants Directive within the UK. In the end, the UK acceded partly because of the need to clarify the financial commitments of the electricity supply industry prior to its privatisation (Boehmer-Christiansen and Skea 1991: 220) (see Plate 10.1).

The domestic agenda in the 1980s: dissatisfactions with traditional approaches

European pressure was the major theme of the 1980s. But European pressure interacted with complex domestic debates in the UK, strengthening the position of those who advocated a more positive environmental policy and who pressed for the reform of pollution control arrangements. In addition, and somewhat paradoxically, the deregulation agenda of the Thatcher years played a role in stimulating change.

The Department of the Environment (DoE) spent much of the 1980s attempting to raise the profile of environmental policy. This task was greatly eased after the 1988 Thatcher speech to the Royal Society which highlighted environmental challenges. But important achievements had been made even before then. The DoE won control of industrial pollution back from the Health and Safety Executive in 1987 with the creation of Her Majesty's Inspectorate of Pollution (HMIP). This was an amalgamation of several bodies with responsibilities for air pollution, water pollution, radioactive substances and hazardous waste. It provided an institutional home for integrated pollution control, as well as acting as the flagship for a more assured environmental policy, but it lacked a strengthened legal framework within which to operate.

One driver for the creation of HMIP was the desire for a more ambitious and more coherent environmental policy. But there were other contributory elements, namely the link to deregulation and, perhaps surprisingly, the support of industrial interests. One of the most important pieces of legitimisation for HMIP was a 1986 Cabinet Office Efficiency Scrutiny Report on industrial inspection. This recommended the creation of a single body responsible for industrial pollution inspection in the interests of public sector efficiency and the reduction of burdens on industry. HMIP was to be a 'one-stop' regulatory shop as much as an institutional mechanism for implementing an integrated approach to pollution control.

Industrial support for institutional and legal change was more complex. Some companies feared the withering of the traditional UK discretionary approach under pressure from European legislation. There was a belief, perhaps naïve, that strengthened British institutions would act as a bulwark against perceived European threats. But there was also a more sophisticated view. Broadly based organisations such as the Confederation of British Industry supported the creation of HMIP because it felt that British pollution control institutions had become so weakened by neglect that they no longer served industry's need for credible regulatory control that would ensure public legitimacy.

The final domestic component of IPC came with the passage of the 1990 Environmental Protection Act. This was an omnibus Act with measures covering air pollution, waste, water and litter. In one fell swoop, UK legislation caught up with the pressures and developments of the 1980s:

- The best practicable means (BPM) principle was substituted by that of *best available techniques not entailing excessive cost* (UK-BATNEEC). This drew on the 1984 Air Framework Directive but substituted the term *techniques* for *technology*, reflecting British concern about practice and procedure.
- The *best practicable environmental option* (BPEO) had to be considered for processes where cross-media environmental impacts were probable. BPEO was propounded by the Royal Commission on Environmental Pollution in 1988 as a means of implementing the integrated pollution control philosophy.
- An authorisation-based approach, with authorisations being reviewed every four years, was adopted, meeting the gradual adaptation requirements of the 1984 Air Framework Directive.
- Public access to regulatory information and emission monitoring data was provided for.

Industry was brought into the new IPC regime between 1991 and 1996. Initially, it appeared that IPC might mark the culmination, as suggested by Table 15.1, of a decade-long process through which European formalism triumphed over traditional British pragmatism and site-level discretion. The following sections show how this has proved not to be the case.

IMPLEMENTING INTEGRATED POLLUTION CONTROL (IPC)

The IPC legal framework is more formal than its air pollution predecessor. Operators must submit an application demonstrating BATNEEC in order to obtain process authorisation. HMIP then issues an authorisation containing legally binding conditions of operation. Operators periodically submit monitoring results to demonstrate compliance. These documents are available on a public register.

HMIP initially wanted to implement IPC in a formal and transparent manner, maintaining an 'arm's length' relationship with industry (HMIP 1990: 17). Inspectors would judge operators' applications against centrally set standards, based on criteria relevant to the industrial sector as a whole and not those for individual sites (DoE 1993a: 12–13). There would be more emphasis on formal emission limits and less scope for the exercise of judgement by individual inspectors. Several factors led to the adoption of the new approach. First, HMIP's top leaders, drawn from the Radiochemical

Table 15.1 Britain and Europe: developments in pollution control, 1976–2007

	UK	EU
1976	Royal Commission Fifth Report: *Air Pollution: An Integrated Approach*	
1982	Royal Commission Fifth Report recommendations rejected by government	
1984		Air Framework Directive
1986	Efficiency Scrutiny Report: *Industrial Inspection*	
1987	HM Inspectorate of Pollution established	
1988	Royal Commission Twelfth Report: *Best Practicable Environmental Option*	Large Combustion Plant Directive
1990	Environmental Protection Act	
1991	Integrated Pollution Control system introduced	Consultations on Integrated Pollution Prevention and Control (IPPC) Directive
1991–93	Flirtation with 'arms length' approach	
1993	Consultative regulatory style re-adopted	Draft IPPC Directive published
1995	Environment Act	Council of Ministers agrees IPPC
1996	Environment Agency established	European Parliament delivers opinion on IPPC. Council of Ministers adopts
1999		IPPC compliance deadline. New BAT limits for large combustion plant take effect?
1999–2007	New plants enter IPPC	
2003		Current Large Combustion Plant Directive expires
2005		Revised large combustion plant national quotas begin to take effect?
2007	Deadline for existing plants entering IPPC	

Inspectorate, were familiar with an arm's length approach (Coleman 1992: 6). Second, the Air Framework Directive suggested that the cost criteria for BATNEEC should be sector-related rather than site-specific. Finally, the DoE wanted the advent of HMIP and IPC to signal a 'cultural break' with the previous air pollution regime.

The elaboration of BATNEEC standards through central HMIP guidance became an important element of the new approach. Like its predecessors, the BPM Note, 'Chief Inspector's Guidance Notes' contained limits and standards which HMIP believed to represent BATNEEC.[2] Under the arm's length regime, the role of Guidance Notes was enhanced, providing prescriptive, quasi-uniform standards and emission limits which all new processes had to meet, and to which existing processes would need to upgrade (HMIP 1992: 6). In this respect, IPC appeared to follow EC norms.

In practice, IPC implementation soon reverted to traditional British practice. The new approach had not taken account of the fact that a solid information base was required for the authoritative setting of BATNEEC standards. The only sources of information concerning the technical, environmental and economic aspects of processes and pollution control techniques were operators themselves. The informality of the previous air pollution regime, plus the large number of processes new to regulation, meant that in many instances no quantitative data were available to HMIP.

Moreover, industry was concerned about being excluded from the operationalisation of the new BATNEEC principle and challenged BATNEEC standards in draft Guidance Notes. In reaction, HMIP began emphasising that standards in Guidance Notes were non-prescriptive (HMIP 1993: 1). In May 1993, HMIP officially announced the abandonment of its arm's length approach. Industry now participates more deeply and much earlier in the development of HMIP regulatory guidance (ENDS 1996b: 34).

Other factors contributed to HMIP's re-think. Experience began to suggest that inspector–operator negotiations at the site level were unavoidable. Just as an absence of information had stalled centralised standard setting in the preparation of sector-level Guidance Notes, a similar lack of information in process applications made it difficult for inspectors to set BATNEEC standards for individual sites. Operators were unfamiliar with the new regime and some had limited competence in environmental management. As a result poor quality applications were submitted, with less than a quarter of operators providing their own BATNEEC assessments and even fewer attempting to determine the BPEO (Smith 1996; Allott 1994).

Staffing constraints and a tight implementation timetable meant that HMIP was unable to make a wholesale request for more information from operators. Nor could inspectors impose stringent authorisation conditions for fear of generating large numbers of appeals, which would have greatly delayed implementation. At the same time, IPC was being singled out for criticism under the government's deregulation drive, which was itself strongly influenced by industry (Business Deregulation Task Forces 1994: 30–33). This symbolised a lack of wider political support for HMIP in carrying out its implementation task.

A multiplicity of factors – scant regulatory information, organisational constraints, industry opposition, and lack of political support – therefore forced HMIP to abandon its arm's length approach. The return to a more

273

co-operative relationship with industry appears to have been endorsed by the new Environment Agency which took over HMIP's responsibilities from April 1996. The result has been a return to the site-specific flexibility and close operator participation which characterised the former air pollution regime. Many authorisation conditions take the form of a request for operators to undertake information-gathering tasks, such as release monitoring, which should have been carried out in preparation for submitting an application (Merrill 1994). Many processes have been authorised with release limits set at emission levels reported by the operators. The National Rivers Authority limits for discharges to water have been simply 'grandfathered' into IPC authorisation conditions. There has been almost no balancing of releases to different media.

Furthermore, the process of negotiating BATNEEC improvements has once again become private to inspectors and operators. The basis of the negotiations between them and the rationale behind any ensuing standards and improvements are hidden; only the outcome of consensus decisions will reach the public register.[3] Within the formal IPC legal framework, there persists an informal regulatory regime.

INTEGRATED POLLUTION PREVENTION AND CONTROL

The adoption of IPPC in Europe

The European Commission began to prepare the IPPC Directive in 1991, just as the UK IPC system began to operate. A formal proposal was introduced in September 1993 (CEC 1993b), with much of the groundwork having been conducted by a UK expert on secondment to Brussels. The basic environmental objective of the proposal was 'to prevent or solve pollution problems rather than transferring them from one part of the environment to another'. The Commission drew on recommendations made by the OECD Council in January 1991 to support its proposal. A key aim was

> to prevent or minimise the risk of harm to the environment taken as a whole; in other words to arrive at the 'best environmental option' which prevents the emission of potentially polluting substances wherever it is practicable to do so, or minimise such emission where it is not.

This phrasing draws heavily on Section 7 of the UK Environmental Protection Act.

Both IPC in the UK and IPPC in the EU can be seen as part of a broader movement towards 'integrated' systems of pollution control among industrialised countries. Why was there a need to take action at the European level? The main justification was that, unless the EU acted, 'Member States having already introduced an integrated approach would be hindered by the existing media-oriented Community legislation to obtain the full environmental benefit of their initiative' (CEC 1993b).

The Commission also offered an 'economic efficiency' justification for the measure: 'taking account of all releases together is likely to be less costly to an industrialist than requiring him to add on technologies or measures to deal with releases to each environmental medium separately' (CEC 1993b). In order to promote a more preventive approach to the environmental management of large, complex industrial processes, Article 2a of the Directive requires that member states take account of a number of factors in granting authorisations to plant operators:

- operators should follow the well-known 'waste ladder' principle – waste avoidance should be given priority over waste recovery which, in turn, should have priority over waste disposal (see Chapter 11);
- energy should be used efficiently;
- measures should be taken to prevent accidents and minimise the impacts of any that do occur; and
- measures should be taken to avoid pollution risks when industrial activities cease – addressing prospective issues of contaminated land and plant decommissioning.

Key issues in integrated pollution prevention and control

Negotiations on the draft Directive proved to be protracted. This was perhaps not surprising, given the diversity of approaches taken at the national level. Individual countries were keen that the European system should conform as closely as possible to their own approaches. A handful of issues that emerged during the negotiation of the Directive raised very basic questions that went far beyond those of administrative compatibility.

First was the concept of Best Available Techniques (BAT) which is at the heart of the IPPC Directive. Classified installations can be operated only with a permit which takes into account the BAT principle. The definition of BAT represents a compromise between two poles in terms of national approaches – the German *best available technology* approach and the UK *best available techniques not entailing excessive costs* philosophy. Although BAT was chosen, the underlying definition owes much to the British approach. As already mentioned, the use of the term 'technique' symbolises a British concern with management and maintenance as well as the technology itself. The definition of 'available' explicitly mentions economic and technical viability and notes the need to consider 'costs and advantages'. British concerns about 'excessive costs' are thus subsumed in the definition of availability. Moreover, any BAT permit conditions must take into account 'the technical characteristics of the plant concerned, its geographical location and the local environmental conditions' (Article 8.2a).

Second, the degree to and manner in which IPPC depends on emission limits was another issue at stake. The final draft Directive (CEC 1995a) signals more flexibility for individual member states than did the initial proposal. Table 15.2 shows key ways in which the text of the Directive evolved. Under the original proposal, it would have been possible to imagine the European Commission systematically collecting BAT information from member states, synthesising it and issuing daughter Directives which would specify emission limits that would constitute a performance floor for permitting arrangements at the member state level. The final draft addresses the question of EU emission limits much more explicitly, but also bounds the circumstances in which Community-level limits might be used. Any requirements would have to emerge from systematic exchanges of information, to be published every three years, and the Council can act only if the

Table 15.2 The IPPC Directive: the evolving approach to emission limits

Original proposal	Final draft
Permit shall normally include emission limit values for substances, other than those emitted in trace amounts and which cannot cause pollution	Permit shall include emission limit values of pollutants emitted in significant quantities, having regard to their nature, the installation concerned and requirements to prevent the transfer of pollution from one medium to another
In certain cases, emission limit values may be supplemented or replaced by other equivalent parameters	Where appropriate, limits may be supplemented or replaced by equivalent parameters or technical measures
Limit shall be based on the best available techniques and shall at least meet those set at the Community level	Limits shall be based on the best available techniques
The competent authority will take into account information on the best available techniques made available by the Commission	
	Acting on a proposal from the Commission, the Council shall set emission limit values for classified installations and polluting substances for which Community action is required on the basis of the exchange of information

Commission proposes.[4] UK policy makers are adamant that this does not 'set up a system of Community emission limit values' but the Directive could be used for such a purpose if the political climate allowed it. In general, the new requirement to take account of the 'installation concerned' in setting emission limit values shifts the locus of rule making away from technical committees in Brussels down towards traditional UK site-level discretion. However, the Directive allows scope for this battle to continue.

The third issue concerns the deregulatory goal of one-stop industrial permitting that is only partly met by the IPPC Directive. Even the original draft allowed for the possibility of several competent regulatory authorities, each presumably covering a different environmental medium. However, the original draft would have required the appointment of a single 'supreme authority' to co-ordinate licensing procedures. As it turns out, member states will only have to ensure that procedures and conditions are 'fully co-ordinated' – the use of separate permits from bodies responsible for different environmental media will remain possible. This obviates the need for institutional reform where integrated permitting is not practised.

It will also not be necessary for a permit to cover all aspects of operations. Member states will be able to formulate 'general binding rules', as long as these provide for an 'integrated approach' and an 'equivalent high level of environmental protection'.

Fourth, from a practical rather than a philosophical point of view, the list of process categories to be covered by the Directive was a critical area of debate – especially for the UK. The draft agreed by the Council of Ministers includes several process categories which the UK Government and sections of industry did not want included. These include slaughterhouses, certain food processing plants and intensive livestock rearing facilities. The UK Government rather explicitly invited those affected to lobby the European Parliament to remove the processes that it had failed to have deleted at the Council table (DoE 1995a).

Fifth, the relationship between BAT, environmental quality standards and emission limits proved extremely contentious. The original draft proposed that BAT standards could be relaxed as long as environmental quality standards or WHO guidelines were being met. This would have served the interests of some Southern member states, especially Spain. This possibility was removed, with UK support, from the final draft. However, additional measures (exceeding BAT) must be used if environmental quality standards are not being met. This provision is not liked by many sections of European industry which would prefer the retention of a more uniform technology-based approach, avoiding the need for detailed site-by-site environmental assessments.

Finally, the speed at which existing processes are drawn into the IPPC system and the frequency with which permits are reviewed is an area in which the UK demands more than will be required under the new European

Directive. The UK requires authorisations to be reviewed every four years, while the original IPPC proposal was for a ten-year review cycle. A compromise of eight years was discussed during negotiations. In the end, the issue was left to member states to decide – competent authorities will need to reconsider permit conditions 'periodically' or if there are changes in pollution levels, in BAT or in Community or national legislation. This will effectively allow each member state to carry on as before.

THE UK, EUROPE AND POLLUTION CONTROL

The UK's role in developing IPPC

The UK played a much more constructive role in the development of the European IPPC regime than has been the case for many Directives dealing with pollution problems. Is it reasonable to deduce therefore that European influence over British environmental regulation has diminished? That British practices have begun to affect European environmental law more fundamentally? Or that, in the context of debate about the application of the subsidiarity principle, UK–European relationships have entered a new phase?

Some specific features of the IPPC debate help to explain the degree of UK influence. First, the IPPC Directive is concerned primarily with environmental procedures (how and under what conditions authorisations for industrial sites may be granted) rather than environmental objectives. The UK has a greater degree of competence and is more comfortable with procedural measures than with those that make specific technical demands. The Regulation on the Eco-Management and Audit Scheme (EMAS 1836/93) is another example where the UK was able to play a constructive role, working on the basis of procedures which had been developed under British Standard 7750.

Second, the UK found itself quite close to the 'centre of gravity' in the IPPC negotiations. On the one hand, countries such as Germany were in a familiar position, arguing that IPPC should provide for a more uniform and formal technology-based approach to standard setting. On the other hand, countries such as Spain were arguing that BAT standards could be relaxed as long as environmental quality standards were being met. By falling mid-way between these positions, UK negotiators were relatively comfortable with the common position which was finally adopted. In addition, UK negotiators were more flexible than has often been the case in the past, being prepared to support the adoption of a measure which did not satisfy all their desires, but which took account of key concerns.

Finally, the fact that the UK had recently enacted legislation implementing an integrated pollution control regime at home was a critical factor. The European Commission frequently relies on available national models when developing new legislative proposals. The British system provided one of the most up-to-date and relevant national models. That the Commission brought in a seconded UK expert to help to draft the Directive both reflected and reinforced this fact.

European impacts on UK practice

Taking a slightly longer view, the impact of European policy on the UK has been considerable, notwithstanding British influence in the specific case of IPPC. The UK IPC regime would not have been instituted had pressures to implement European Directives not created a need for new primary legislation. During the 1980s, various bodies – notably the Royal Commission on Environmental Pollution and sections of the business community – supported the revision of the UK pollution control regime. But at least some of this support can be seen as a reaction to European pressure, and on its own it would probably have been insufficient to instigate legislative change. European factors were critical in triggering a forward movement in UK policy.

The practical consequences of this pressure in terms of site-level implementation are somewhat mixed. For a time, as HMIP pursued its arm's length regulatory policy, it appeared that the formal 'European' approach had triumphed entirely. Even without this, the British system now has many more formal elements than was the case fifteen years ago. The BATNEEC principle is interpreted in a more specific and explicit way in authorisation conditions. More use is now made of formal emission limits, though these may vary from one site to another. Information on authorisation conditions and emission monitoring is now available in the public domain through public registers, making the system more transparent and accountable. On the other hand, the collapse of the arm's length policy has seen a return to the traditional British approach of site-level discretion exercised by individual regulators. In practice, much regulatory negotiation is taking place once again in unrecorded meetings between industry and HMIP.

Experience with IPC implementation suggests IPPC may have specific impacts on UK practice, but in other respects little will change. IPC and IPPC employ similar regulatory principles with the same site-specific caveats, 'thus preserving the UK's site-specific approach to authorisations' (DoE 1995a). IPPC does not alter the inherent asymmetry of regulatory information available to operators and inspectors under any technology-based regulatory system (Weale 1992: 177; Davies and Davies 1975: 227). IPPC implementation is likely to follow the IPC approach: inspector–operator negotiation of consensus-based, site-specific regulatory goals.

IPPC could increase the number of processes regulated at the national level, though the precise number is unclear. The Directive embraces industrial sectors which are not currently subject to IPC; and IPPC covers more substances than the twenty-three prescribed under IPC. It has been calculated that this will add as many as 5,000 processes to the 3,000 covered by IPC (ENDS 1995c: 38). Yet the DoE believes it can avoid bringing new processes into IPC (DoE 1995a).

Article 8.6 of the IPPC Directive allows regulators to set general binding rules for the integrated control of releases for a whole class of processes rather than requiring individual authorisations. The DoE 'will be looking into the extent to which it can make use of this flexibility for some of those processes which do not currently need an IPC permit' (DoE 1995a). More-over, processes covered by existing, single-medium controls do not need to be authorised so long as they have a combined effect equivalent to IPPC. It is debatable how many discharges currently covered by the UK's water pol-lution regulations, based upon environmental quality criteria, are compatible with the technology-based BAT principle of IPPC. Any increase in the number of processes regulated has resource implications for HMIP. New Guidance Notes might need to be prepared. More processes would require authorisation and inspection.

The 1990 Environmental Protection Act promised little in the way of formal attention to preventive approaches to the control of pollution. How-ever, the 1995 Environment Act is more explicit in requiring the new Environment Agency to 'promote sustainable development'. The Agency, following up on initiatives begun by HMIP, will encourage operators to take more preventive action by promoting waste minimisation projects and per-haps reducing authorisation fees for companies which have adopted suitably certified, environmental management systems. The 1995 Act and the approaches being adopted by the Agency are likely to be sufficient for com-pliance with the requirements of IPPC with respect to preventive approaches.

Overall, IPPC is unlikely to affect the UK's basic approach to regulating industrial processes. There is scope in the Directive for Europe-wide, uni-form emission standards on the basis of a qualified majority vote. But this is enabled by the IPPC Directive rather than being mandated. There is little likelihood of uniform BAT standards being introduced across a range of processes. However, traditional battles over emission limits are unlikely to disappear. The European Commission has recently drafted a revision, for consultation, of the controversial Large Combustion Plants (LCP) Direct-ive. The UK may choose to question the need for this measure, given the scope for action at the member state level under IPPC.

THE UK AND EUROPEAN ENVIRONMENTAL POLICY: A NEW PHASE?

Apart from the specific features relating to the development of IPPC described above, there are reasons to believe that UK influence over Euro-pean environmental policy will be stronger than in the past and may be entering a new phase. First, the widening of the membership of the Community in the 1980s means that the UK is now no longer among the most reluctant member states when new environmental measures are proposed. Spain, for example, generally has more difficulties than the UK. However, the recent accession of Sweden, Austria and Finland may strengthen the Community's 'green bloc'. Second, the flavour of Community environmental policy itself is changing, as reflected in the Fifth Environmental Action Programme. There is now a greater emphasis on 'social actors', policy process and procedural matters. This is likely to strengthen the hand of the UK which demonstrably has a national competence in designing and implementing procedural measures. Finally, rising and widening concern about 'subsidiarity' is pushing real decision making power further down the implementation chain. This matches traditional UK approaches which emphasise local discretion and negotiated solutions.

ACKNOWLEDGEMENTS

Jim Skea is grateful to British Gas plc and the Economic and Social Research Council for support for this work. Adrian Smith has just completed a DPhil project on the implementation of integrated pollution control in the organic chemicals industry. All the views expressed in this chapter are those of the authors alone.

NOTES

1 Pollution control in England and Wales was the responsibility of HM Inspectorate of Pollution until March 1996. The Inspectorate, along with the National Rivers Authority, was absorbed into the Environment Agency in April 1996.
2 The formal role of a Chief Inspector's Guidance Note is to provide guidance to individual inspectors in their dealings with operators. The Notes are published and hence appear to have a somewhat wider significance.
3 Until April 1996 even operator reports on the feasibility of upgrading to BAT-NEEC standards were not placed on the public registers, but since then this regulatory anomaly has been corrected.
4 In fact this only re-states the formal position under the Treaty of Rome.

Part V

CONCLUSIONS

Part V

16

LESSONS AND PROSPECTS

The prospects for the UK environment in Europe

Philip Lowe and Stephen Ward

From the evidence of the different chapters it is clear that the responses to European integration have been as varied across the environmental sector as they have been pervasive. Perhaps the only sure conclusion to be drawn therefore is that British environmental politics and policy have been profoundly affected by European integration. The separate organisations and policy fields provide distinct perspectives on the process of Europeanisation, and one must be wary of over-generalisations from any one of them. Some general lessons, though, can be drawn by examining experiences across the board. This chapter does so by concentrating on the three major areas addressed in the book: Government–EU environmental relations; policy and policy making; and domestic institutional restructuring.

GOVERNMENT–EU ENVIRONMENTAL RELATIONS

Despite the rise of environmental issues in national and international politics over the past decades, the environment has never been a central concern of Britain's relations with the EU. This reflects the secondary status of EU environmental policy and is also indicative of UK governments' treatment of the environment as a minor policy sector. Overall, Britain's environmental relations with the EU have been shaped by three main factors: the predominant political outlook towards Europe; the type of issue under negotiation; and the attitude and leadership (or lack of it) of the responsible Minister.

The predominant political view of the EC in Britain has been that of it as a single market. The overriding objective has been that of trade liberalisation. There has been little interest in, but considerable antipathy towards, political as opposed to economic integration. The growing ramifications of integration beyond trade and tariff reforms have therefore provoked a mounting political backlash.

With this overall orientation, UK governments have tended not to regard EU environmental policy as a sector for proactive initiative. Often, instead, their reactions to developments in this sector have reflected the economic and industrial priorities behind Britain's membership of the EU. Robin Sharp (Chapter 2) characterises Britain's stance as 'pragmatic ... keen to ensure that environmental measures take full account of other interests and are not unreasonably expensive'. He provides a number of examples where the economic Departments in Whitehall determined the UK's negotiating stance on key environmental matters – for example, in ensuring the discretion of member states to allow development in Special Protection Areas and Special Conservation Areas; over Community co-financing of national measures taken under the Habitats Directive; and on the exemption of farming and the food industry from the Integrated Pollution Prevention and Control Directive. As Edwin Thairs (Chapter 9) explains, 'It was expected that the British government would look after the interests of British industry in Brussels negotiations.'

This does not mean that Britain has lacked an EU environmental agenda. In many respects, though, the point of departure for that agenda has been the wider political objectives of Britain's membership: for example, promoting a level playing-field for industry, safeguarding British sovereignty, combating over-regulation and supporting environmental measures that are cost effective and economically feasible. Such concerns are echoed in Britain's approach to other EU policy sectors.

The environmental sector does differ, however, in three important respects, all of which have made the acceptance of a European framework of policy and regulation less contentious. First, the rationale for supranational action is often easier to justify because of the evident transnational nature of the most pressing environmental problems and the demonstrable failure of nation states to tackle them. Second and relatedly, there are consistently high levels of public support for EU action in this field, unlike for many other policy areas (see Figure 1.3). Third, in the domestic arena the environment has never emerged as a focus of conflict between the major parties.

In such circumstances, issues could be displaced to the European level without Ministers courting unpopularity or a political backlash. The Europeanisation of environmental policy has not caused the sort of contention between the UK parties that, say, social or monetary policy has where existing partisan differences have been inflamed by Euroscepticism. Of course, implicit in these contrasting responses are judgements of the political significance of different issues. Domestic control over environmental standards clearly ranks well below control of the currency or interest rates or the armed forces or the country's borders in most politicians' considerations of the strategic national interest. Such a judgement, coupled with the strong popular support for EU action in the environmental field, has led to a certain tendency to treat the environment as a 'sacrifice issue' in Britain's crabwise progression towards European integration. A positive stance on EU environmental policy has been adopted by the major parties in the domestic arena and by UK governments in the European arena, to offset the negative stance adopted in other policy fields considered more vital to the national interest.

The British vision of the EU as an intergovernmental partnership of nation states with priority to free trade issues remains the predominant one in Europe, and UK governments have been obliged to accept the enlarged role for the EU in environmental policy making which other member states have pressed for as the counterpart to market liberalisation. Britain's interest in the development of EU environmental policy has thus been in the consistency of standards within the Single Market and in accommodating its domestic procedures to the development of EU rules. However, these two concerns can pull in different directions, with demands for more subsidiarity and flexibility in EU legislation on the one hand, yet on the other calls for more consistent implementation and compliance across the member states to promote a level playing-field. These seemingly contradictory positions tend to be deployed in different contexts: with level playing-field arguments used where the environmental issues concern British industrial competitiveness in international and European markets (typically over product standards and with the DTI playing an active, if not the leading, role); and flexibility arguments when the issues concern domestic environmental management (for example, water quality or waste management, more the strict preserve of the DoE).

The approach towards EU negotiations thus varies according to the subject. Given the fragmented institutional structure of the EU and the wide variety of detailed, often rather technical issues that it covers, sectoral Ministries have a relatively big say in EU politics, not least through their direct input into the sectoral Councils, Council working groups and expert groups. By and large, the UK Department of Environment and its Ministers have not seen the EU as an avenue for pursuing high environmental standards. Britain's main contributions to EU environmental policy, indeed, have been procedural rather than substantive (such as waste management plans, environmental auditing and integrated pollution control), reflecting a national tradition in which central policy makers are used to laying down the procedures and structures for regulation but leaving to local implementation and discretion the outcomes to be achieved thereby. Arguably, much of the rationale of British negotiations in the EU has been to preserve that procedural logic, often defensively but occasionally offensively (by promoting British procedural models in the formulation of new directives).

Sharp's chapter also points out that the leadership of individual Secretaries of State and their attitudes towards European integration can make a difference. It is interesting to note that the UK's most difficult period with the Community over environmental policy came in the time of Nicholas Ridley (an arch Eurosceptic) at the DoE. Similarly the deregulation and subsidiarity agenda came to the forefront in 1992 during the brief tenure at the DoE of Michael Howard, another Eurosceptic. In contrast, during the final years of the Major Government, when Britain's relations with the EU came under increasing strain, the Secretary of State for the Environment was John Gummer, a notable Euro-enthusiast. Under his leadership, the DoE developed a European orientation and Britain came to play a more active and positive role in the development of EU policy. Of course, where environmental issues spill over into other sectors, the scope for exercising such initiative may be curtailed, particularly when more powerful Whitehall Ministries are involved (such as the Treasury or the DTI) or ones more central to the UK–EU axis (such as MAFF or the Foreign Office).

Where does this leave the UK's reputation as 'the Dirty Man of Europe'? The reputation was earned in the 1980s more for the country's standing as an awkward partner than for its objective environmental performance, although Britain's contribution to acid rain in Scandinavia and the radioactive contamination of the Irish Sea are nothing to be proud of. More generally, though, the reputation reflected the UK's defensive and unconstructive response to the European environmental agenda, the difficulties of adapting existing domestic structures to the new EC legislative frameworks, and the UK environmental lobby's active use of European channels to challenge its own government, resulting in a number of high-profile clashes between Britain and the Commission. Certainly the chapters in this volume indicate that, although Britain has not often exercised progressive leadership, it has not, in the main, been a lag state over EU environmental policy. After a period of catching up (in the late 1980s and early 1990s) and the convergence of domestic and European agendas (in the mid-1990s), Britain could more accurately be categorised as a middle-ranking member state (see Chapter 2). Its set of preoccupations often puts it at odds with 'green' member states such as Denmark, Germany and the Netherlands and tends to place it in a group, including France and Spain, which shows a pragmatic interest in the development of EU environmental policy. This group usually strongly tempers the pursuit of common and improved standards with concerns for industrial competitiveness and minimising the burden on their own economies and national systems of implementation (see Chapter 4).

ENVIRONMENTAL POLICY: STYLE AND SUBSTANCE

In terms of policy, this book has concentrated on three major themes: how has the style of policy making been altered by Europeanisation? what determines the distinctive patterns and pace of integration in different fields? and what have been the consequences for the substance of policy and the environmental standards that apply?

There have undoubtedly been significant changes in the procedures and principles of environmental policy. European Directives have required that absolute legal standards be put in place for a range of environmental protection parameters. This has involved a shift from flexibility to formality in formulating and implementing the objectives of environmental policy. A common European framework, coupled with legal standards to be met over proscribed timetables, has greatly reduced the scope for discretion in implementation and has helped create a more transparent system that is much more open to public and judicial scrutiny. This has necessitated making explicit the principles upon which environmental protection is based, and increasingly the justifications for policy have drawn on concepts developed in European spheres, such as the primacy of law in regulation, the notion of subsidiarity, the principle of precautionary action and the polluter pays principle.

Clearly, these are profound changes, but before pronouncing a transformation in Britain's traditional policy style, certain caveats need to be entered. On the one hand, Ward (Chapter 14) challenges the notion of a general tradition in British environmental policy. In contrast to the pragmatic incrementalism claimed in conventional accounts (see Chapter 1), he points out that water management, at least, has long been organised around a very clear, logical and strategic framework, albeit one pervaded by technocratic rather than legal norms. On the other hand, aspects of Britain's traditional approach remain very much apparent. Jordan (Chapter 10) singles out the following features: a cleaving still to secrecy in environmental regulation; the continued prevalence of administrative discretion over judicial interpretation; and a strong attachment to informal voluntary agreements and unquantified standards. Certain practices have proved remarkably resilient, none more so than the site-level regulation of industrial pollution which means, as Skea and Smith point out (Chapter 15), that detailed and secretive negotiations between the polluter and the pollution inspector remain at the heart of industrial pollution control, despite the generalised claims for a more transparent and codified approach. Finally, the changes that have occurred are not all due to European integration. Another key factor has been the growth in public concern for the environment linked to the professionalisation of the environmental lobby (Chapters 5 and 6) and the explosion of information on the state of the environment (see Chapter 7). Domestic developments not specific to the environmental sector have also been important, the most pervasive being the restructuring of public administration which, in the case of the privatisation of utilities, for example, has necessitated new forms of regulation in fields such as water quality (Chapter 14) and waste management (Chapter 11). Moreover, quite apart from the Europeanisation of environmental policy, there has been an internationalisation of the environmental agenda which has obliged nation states to come to terms with transnational and global environmental problems. European integration has interacted with these other factors with consequences that differ significantly between countries (see Buller's comparison of the UK and France, Chapter 4) and between policy fields (see below). It would be a mistake, therefore, to presume that a national style of environmental policy is being replaced by a European one.

How, more specifically, can we understand the distinctive pattern and pace of integration in different policy fields? – the second of our policy themes. The fields we have examined could be arranged from land-use planning, as the least Europeanised, to water quality, as the most Europeanised, with nature conservation, waste management and industrial pollution control placed in between. This reflects the different volume of EU legislation in each field, with very little legislation in the land-use planning field and a detailed legislative framework for water quality. The pace of integration also reflects the differential progress of EU environmental policy. The fields of water and air pollution were developed in the 1970s and extended in the 1980s. Waste management, though also addressed at the EC level in the 1970s, was the subject then of broad and very general framework Directives with limited consequences nationally, and was reactivated as a policy field in the mid-1980s. The Europeanisation of nature conservation has been a much more limited and recent affair which started with the 1979 Birds Directive but only really took off with the 1992 Habitats Directive and Agri-environment Regulation.

The dynamics of UK–EU integration in specific fields has reflected the policy distance to be bridged. Waste management, for example, was just being established as a policy field when the UK joined the EC: in consequence, there was considerable scope for the UK both to exert influence upon, and to respond to, the emerging EC framework for regulating waste. In contrast, water policy was a mature and well-established field with its own clear and logical framework which had developed in isolation from the Continental experience of joint management of common rivers, lakes and watersheds. Unsurprisingly, there was a considerable gulf in the policy and procedural norms involved, and the adaptation of British practice to a European framework has been a protracted and contentious process.

Of course, EU regulatory concepts and approaches do not emerge from thin air. It has been argued that EU policy is, in fact, a patchwork of national regulatory models – the product of 'regulatory competition' between governments seeking not only to shape the EU agenda of problems to be processed but also the basic regulatory philosophy and the policy instruments to be applied (Héritier 1992; Héritier *et al.* 1994). In the preparatory phase of EU legislation, the Commission typically looks to the member states for viable strategies and solutions to be transferred to the European level, often via seconded experts (Pellegrom 1997). States wishing to retain their regulatory cultures and problem-solving approaches seek to impose their own regulatory models on the EU. Of course, for states to be able to take the lead they must have relevant models in place that fit in with the Commission's overall policy objectives and are capable of recruiting sufficient political backing from other member states. Anderson and Liefferink suggest that 'member states, which have been innovators or forerunners on a specific issue, can support the development of EU policies on the basis of their domestic experiences' (1997, chapter 1). In the early 1980s, for example, Germany was successful in promoting its emission-oriented air pollution control policy at the EC level. The UK, with a different traditional approach and lower consciousness of transboundary air pollution in general and acid rain in particular, resisted this development. Thus, through the 1980s, the UK was out of step with the development of EU air pollution policy, particularly over the Large Combustion Plants Directive and the Directives on vehicle emissions. It did take the opportunity, however, to renew its national approach to air pollution and from the late 1980s onwards has sought to take the offensive in seeking to promote its own innovations, such as integrated pollution control, as the model for EC policy development (see Chapter 15).

A review of the experience in the policy fields covered in this book suggests the significance of policy cycles in connecting domestic to EU policy processes. In fields where innovation has recently taken place in the domestic sphere, policy makers are encouraged to push their approaches at the European level too. They are able to do so to the extent to which they are part of emergent European policy networks, involvement in which also allows them to anticipate EU developments. Waste management policy provides an instructive example. Porter (Chapter 11) identifies 'periods in early and more recent policy where the UK has been influential in setting the EU agenda and providing useful models for EU policy'. These periods of British leadership followed immediately after major pieces of domestic legislation – the 1974 Control of Pollution Act and the 1990 Environmental Protection Act – which comprehensively overhauled British practice and provided the basis for an activist approach towards EU policy development. During the period in-between, the 1980s, when British policy lost its momentum amid public expenditure squeezes and political battles between central and local government, the EU agenda was determined by considerations of other member states with different concerns, to which the UK was forced to react.

The third policy theme of the book concerns the consequences of Europeanisation for the substance of policy and the environmental standards that apply. All the policy studies agree that integration has led to higher standards of environmental protection being adopted in the UK. In some cases, notably water policy, the standards have been pushed significantly higher. The obligations to implement EU legislation and to respect agreed targets and timetables have also sharpened up the performance of Britain's system of environmental protection and regulation. Two basic factors account for these improvements. First, within the EU, Britain is harnessed to a common environmental policy which has largely been driven by states with higher standards than the UK that have sought to project those standards on to the EU level. Second, the European framework fosters comparisons between states. It provides common yardsticks and comparative data that allow relative judgements about environmental conditions and the effectiveness of national organisations in different countries. The conventions that formalise the growing transnational flow of environmental information are thus of critical significance (see Chapter 7), not least because the European framework also provides an institutional context in which states must justify the approaches they adopt and come to terms with alternatives. At the very least, officials and Ministers are thereby kept on their toes, but the availability of comparable evidence and alternative models allows others – pressure groups, the media, the public – to draw telling comparisons too. For Britain, the effect in the 1980s was to dispel much of the complacency that had previously surrounded national environmental policy. The charge of Britain being the 'Dirty Man of Europe' stuck, and, though it evidently infuriated Ministers, it spurred them to take action to improve Britain's image.

The environmental improvements achieved have in some cases been extremely costly. For example, £10 billion is being spent between 1989 and 2005 on improving sewage discharges in Britain under the Urban Waste Water Treatment Directive and the Bathing Water Directive. It is inconceivable that we in Britain would have had such an extensive coastal clean-up programme without European action. EU legislation thus drives major investment programmes and indeed overrides domestic priorities. One consequence of responding to an environmental agenda that is largely set elsewhere is that there is little debate about the resultant scale and pattern of national resource allocation and whether this is yielding the optimum benefit for the UK environment.

Moreover, concluding that the influence of Europeanisation is widespread within the environmental sector and that standards have thereby been raised does not mean that European integration has been entirely beneficial for the UK environment; far from it. Benefits gained from the relatively proscribed EU environmental agenda have been offset by the damage arising from other EU policy sectors, including the Common Agricultural Policy, regional policy and the Single Market (see Chapters 6, 12 and 13). The environmental consequences of such economic policies have been extensive and largely detrimental. There have been some initiatives to redress matters, such as the Agri-environment Regulation and the introduction of environmental considerations into the administration of regional policy (the so-called greening of the Structural Funds), but the effects have been marginal. Neither the importance nor the difficulties of integrating environmental objectives into other policy sectors should be underestimated.

DOMESTIC ADMINISTRATIVE AND INSTITUTIONAL RESTRUCTURING

Arguably the most profound consequence of European integration has been the restructuring of the domestic environmental arena. This has encompassed changes in environmental administration; in the roles and rationale of environmental organisations; and in the power relationships between actors in the environmental policy field.

The organisation of environmental administration in the UK has changed considerably over the past twenty-five years. In particular, there has been a movement of powers away from local government to a variety of private sector bodies, agencies and central Departments. European integration has been an important factor but its specific consequences have depended upon the interaction with domestic political forces. This is evident in relation to both of the two major trends, namely centralisation and privatisation.

Turning first to the former, it appears that European integration has encouraged both centralising and decentralising tendencies. The centralising tendencies derive from the transfer of powers to the EU and from the key role of national governments in EU decision making, including their exclusive responsibilities for negotiating Directives through the Council of Ministers and for ensuring their implementation nationally. But there is also a diffusion of governmental power at the EU level through the necessity for states to work through EU institutions and to collaborate with the other member states. EU law also places constraints on executive action. The decentralising tendencies thus reside in the enhanced opportunities for autonomous action for sub-central actors arising from the erosion of the authority of the central state, as well as in the necessity to have appropriate and legitimate sub-central structures to administer EU Regulations and programmes.

According to Buller (Chapter 4), this essential ambivalence explains why the UK, with such strong traditions of devolved administration, has experienced European integration as a centralising force and why France, with strongly centralised traditions, has experienced it as a decentralising force. Both Buller and Morphet (Chapters 4 and 8) suggest that the resolution of these opposing tendencies is expressed through the development of an intermediate tier of regional administration, and that this is the significant structural outcome of Europeanisation.

It is also the case that, through the 1980s, the influences of European integration coincided with governments in the UK that had strongly centralising instincts and an antipathy towards local government. As Sharp and Buller point out, Ministers at the DoE were often too preoccupied with their battles against local government to pay much political attention to the gathering EC environmental agenda. Increasingly they had to respond to that agenda and in doing so they often introduced legislation that curtailed the powers or the autonomy of local government. European policy provided the context and the impetus for these domestic developments but did not necessarily predetermine the specific (centralising) consequences.

To a considerable degree, European integration has played a similarly catalytic role in relation to the other major trend – namely the privatisation of environmental services. Buller (Chapter 4), from the example of water services, suggests that the imposition of mandatory quality norms and standardised regulatory procedures under EU legislation has provided the legitimacy for the entry of private capital into the supply of environmental services, while the additional costs entailed and expertise required have provided the impetus. The comparison between the UK and French experiences indicates, though, that the form, degree and pace of privatisation are determined largely by national factors.

Linked to the reorganisation of environmental administration have been changes in the functions of environmental organisations, including a blurring of previously distinct roles. Not only have private bodies taken over public functions. Governments, local authorities and public agencies have also been obliged to act more as pressure groups in the European arena. In contrast, some of the larger lobbying groups have taken on scientific research and policy development functions once the preserve of government. A striking example is the key transnational role in agri-environmental research, monitoring and policy development played by the RSPB and BirdLife International (see Chapter 12). This blurring of roles arises from the diffusion of power and the erosion of government hierarchies within the EU. In this context, governmental organisations on the one hand are obliged to build coalitions; and on the other, there is a niche for transnational lobbying organisations in filling the information gaps that are a chronic feature of the European policy arena.

These shifts have arisen in part as organisations have tracked the changing locus of power and as a result have come on to the novel political terrain of the EU. A number of chapters recount how various environmental policy actors – local authorities, statutory agencies, pressure groups and business and industrial lobbies – have established their own direct channels into the European arena. Environmental groups were among the first to do so, in response to their marginality in domestic politics, including their exclusion from key policy communities. Business and industrial interests also set up

European co-ordinating structures at an early stage and had their own direct access to the Economic and Social Committee (ECOSOC). However, they did not feel the need to focus on EC-level lobbying on environmental matters as long as the UK Government took a circumscribed view of EC environmental law and saw the main point of European negotiations as being to defend Britain's economic interests and existing domestic procedures.

During the 1980s, as EC environmental law expanded in volume and impinged ever more on domestic procedures, environmental groups were able to develop a role as watchdogs in the implementation of European environmental legislation. Central government found it increasingly difficult to control the flow of environmental information and access to the policy process. With the policy agenda no longer contained within the confines of Whitehall, a number of existing policy communities – such as those on water quality and industrial pollution – were prised open.

Appreciation of the significance of the European agenda led in the late 1980s to a greater emphasis all round on EC lobbying. The bigger environmental groups set up their own Brussels offices, thus bypassing the cumbersome procedures of the European Environmental Bureau (see Chapter 6). But other organisations also realised the need for an effective direct input, including specific industrial and business interests and local authorities (see Chapters 9 and 8). In an increasingly crowded policy field and with growing emphasis on the practicality of EC legislation, there was a premium on relevant technical expertise and implementation experience, which favoured some of these other interests while environmental groups found their influence squeezed (see Chapters 3 and 6).

The continuing centrality of member states in the decision-making structures of the EU and their control over the implementation of European legislation mean that interest groups must maintain their established channels of access to national governments. European lobbying can be no substitute for, but must be additional to, national lobbying (see Chapter 5). The Europeanisation of policy thus establishes as key those individuals and groups that are in a position to connect domestic and European policy processes. Within organisations there has been the emergence of a class of European policy entrepreneurs with contacts and experience in transnational policy networks. Likewise, the organisations that have taken the lead are those that are sufficiently well resourced to maintain transnational lobbying structures. Within different organisational sectors, such as the environmental movement or local government, the effect has been to favour a small elite of groups and authorities, usually those which were already at the forefront domestically. Many other organisations no longer have the capability or motivation to follow the increasingly distant and convoluted policy process from start to finish. The smaller policy actors are restricted to concentrating on implementation politics in the domestic sphere (see Chapter 5).

PROSPECTS: THE UK AND THE FUTURE EUROPEAN ENVIRONMENTAL AGENDA

This book, finalised in 1997, a quarter century after the Treaty for Britain's accession to the Community was signed, provides a timely opportunity to put UK–EU relations into perspective. The election of a Labour government with a landslide majority and Treaty reforms following the 1996/7 Inter-Governmental Conference provide pointers to the future direction of environmental policy. Do these events herald a new phase of UK–EU environmental relations? One which marks a maturing of relations?

Domestic politics and future directions: The Labour Government

The Conservative Party has been in office for twenty of the intervening twenty-five years. It is not unreasonable to expect that a new Labour administration will bring a change of approach. Before examining such expectations it is worth making two cautionary points. First, the Labour Party traditionally has been neither the party of Europe nor of the environment. It is only since the late 1980s that it has fully accepted membership of the EU (Tindale 1992; George and Haythorne 1996) and taken tentative steps towards 'greening' its positions (Carter 1992; Robinson 1992). Second, as noted in the opening chapter, neither European affairs nor environmental policy divides the parties on straightforward ideological lines, but both cut across traditional party divides. Indeed, the UK parties' underlying attitudes towards Europe and environmental policy have not been markedly different. One needs to ask, therefore, whether the change of government will actually produce significant alteration to the infrastructure, principles or substance of UK–EU environmental relations.

Although it is too soon to draw conclusions about the eventual impact of the Labour Government, it has already signalled a number of changes in terms of EU diplomatic relations, environmental policy initiatives and administrative restructuring, all of which could create the basis for a new phase in environmental governance

In general terms, Labour has promised a more constructive and co-operative relationship with the Commission and with other member states. This attempt to mark a fresh start and to distinguish the Labour Government's outlook from previous Conservative administrations' has proved cathartic and has elicited a favourable response from European partners. However, the consequences in the environmental sector are unlikely to be pronounced. As a number of chapters in this volume have noted, the environmental sector has been insulated from the broader difficulties of

UK–EU relations of recent years by the presence of Europhile ministers, the lack of contentious environmental issues, and converging environmental agendas.

As part of this new spirit of co-operation, Labour promised a more flexible stance to proposed reforms emerging from the Inter-Governmental Conference. One notable concession was a willingness to accept a further extension of Qualified Majority Voting (QMV) on environmental matters. While this might seem to signify a new pro-integration line, in reality it reflected broader political considerations. Since in most respects Labour policy on EU institutional reform was not mark-edly different from that of the Conservatives, the environment represented a non-contentious area on which to relinquish sovereignty. Arguably then, this is another instance of the environment being regarded as a sacrifice issue.

In terms of environmental policy initiatives, there are mixed signals emerging from Labour. Theoretically, party policy is still based on the wide-ranging and ambitious document 'In Trust for Tomorrow', produced in 1994 and welcomed by environmental organisations as a major step forward. However, the Labour leadership has avoided unequivocally endorsing the programme. Environment was not a key plank in Labour's election strategy and the specific promises made were both cautious and modest. Labour appeared concerned to discount the impact of any environmental policy commitments on public expenditure or business competitiveness.

However, initiatives immediately following the election may indicate the beginnings of a more adventurous approach. In relation to Europe, two examples are worthy of note. First, the Foreign Secretary's mission state-ment included a commitment to make environmental considerations a key component of foreign policy. This is the counterpart to Labour's domestic commitment to integrate environmental considerations across government. Second, the new government ordered a fundamental review of practices at the root of Britain's 'Dirty Man of Europe' tag, including a reconsideration of policies on the dumping of oil installations in the North Sea, and on radioactive and toxic waste discharges into both the North and Irish seas (*Guardian*, 17 May 1997).

One early administrative reform may herald a more substantive long-term change. The creation of the new 'super ministry' combining environment, transport and regional policy, headed by the Deputy Prime Minister, creates a Department with significant resources and political weight within White-hall. Potentially this could overcome the long-standing criticism of environmentalists that the DoE has been marginalised in the Whitehall hier-archy. Moreover, it also provides the opportunity to 'green' the environ-mentally damaging areas of transport and regional policy, both of which are heavily influenced by EU agendas over which environmental groups have worked at the European level to establish environmental criteria. This organisational reform thus provides opportunities for environmentalists to enter previously closed policy communities in the domestic arena.

There are, then, some encouraging signs. But the test of Labour's pledge to put environment at the heart of government will come in the resolution of the contradictions with more pressing priorities such as economic growth, regional development, job creation and labour market flexibility. Overall, though, one can expect a more positive and proactive approach towards EU environmental policy. Yet here, too, there are broader contradictions. UK–EU environmental relations cannot be segregated from the UK's overall role in the EU, and Labour remains sceptical of further European integration but has yet to provide substantive answers concerning its model of the EU's development.

The EU agenda: towards a flexible Europe

Events in the wider European arena over the next decade may prove uncomfortable but decisive in determining Britain's future in Europe. EMU and enlargement will both deepen and widen European integration. In the short to medium term, the new Amsterdam Treaty provides the framework.

Although the 1996/7 Inter-Governmental Conference represented an opportunity to promote environmental issues, the direct influence of the new Treaty will be limited. Environmental policy was never high on the IGC agenda, being overshadowed by economic and security concerns and the push to enlargement. Since the environment title is relatively new, both the Commission and many member states saw little need for major changes.

Thus the resulting impact of the Amsterdam Treaty in the environmental sector may well lie in institutional and procedural changes, in particular, the extension of majority voting and the bolstering of the powers of the 'greenest' EC institution, the Parliament. The new Treaty will not provide the major symbolic boost that the Single European Act achieved a decade earlier. Its significance will be to refine and consolidate environmental goals. It is the political will of the member states and the priority attached to implementation and integration of environmental policies that remain of greater importance.

One striking theme to emerge from the ICC negotiations was the concept of flexibility. This implies that groups of EU countries co-operate and move forward more rapidly in policy areas where not all member states may wish to take part. Such a concept has been stimulated by a recognition of the need to avoid stagnation in an enlarged EU of twenty or even thirty member states. Indirectly, flexibility raises a number of important issues and dilemmas both for the environmental sector and for UK–EU relations.

In the environmental sector, flexibility would appear practical and beneficial in two situations. First, many prospective EU members face significant pollution problems as a legacy of the Communist era. There appears little hope of many of these states meeting EU environmental standards in the medium term. In principle, flexibility could be used to manage a possibly lengthy transitional period for these countries to absorb the existing corpus of EU environmental legislation. Second, flexibility could offer a way forward where a few member states block agreement by the rest. Thus environmental policy could be pushed forward by progressive member states rather than held back by the slowest. In reality, however, the repercussions of flexibility might be to reduce incentives to co-operate with all the member states and to widen the gap between environmental lead and lag states. In essence, both these scenarios might result in a multi-speed Europe, with different groups of member states working towards similar broad objectives but at a different pace.

Superficially, flexibility might appear attractive to the UK, given its concerns over pooling of sovereignty in sensitive fields. Yet, it has always been wary of being relegated to a European second division, and a resulting inability to influence European affairs. Thus if flexibility were to emerge as a central concept, then the UK would face the ultimate test of its interest in being an influential EU member. In the environmental sector, the UK would have no alternative but to cast off its middle-ranking status and develop a leading, proactive position.

BIBLIOGRAPHY

Note. Directives of the European Union and Regulations of the Council of Ministers are listed in the Index.

Aguilar-Fernández, S. (1994) 'Spanish pollution control policy and the challenge of the EU', in Baker, S., Milton, K. and Yearley, S. (eds), *Protecting the Periphery: Environmental Policy in the Peripheral Regions of the EU*, Ilford: Frank Cass, 102–17.

Alkali and Clean Air Inspectorate (1974) *Annual Report of the Alkali and Clean Air Inspectorate for 1973*, London: HMSO.

Allott, K. (1994) *Integrated Pollution Control: The First Three Years*, London: Environmental Data Services.

Andersen, M.S. and Liefferink, J.D. (eds) (1997) *European Environmental Policy: The Pioneers*, Manchester: Manchester University Press.

Andersen, S.S. and Burns, T.R. (1992) *Societal Decision Making: Democratic Challenges to State Technocracy*, Aldershot: Dartmouth.

Andersen, S.S. and Eliassen, K. (1990) *The Explosion of European Community Lobbying: The Emergence of a European Political System*, report 316–17, Sandvika: Institute of European Studies, Norwegian School of Management.

Andersen, S.S. and Eliassen, K. (1991) 'European Community lobbying', *European Journal of Political Research*, 20, 2 (September): 173–87.

Andersen, S.S. and Eliassen, K. (eds) (1993) *Making Policy in Europe: The Europeification of National Policy-making*, London: Sage.

Arp, H. (1993) 'Technical regulation and politics: the interplay between economic interests and environmental policy goals in EC car emissions', in Liefferink, J.D., Lowe, P.D. and Mol, A.P.J. (eds) *European Integration and Environmental Policy*, London: Belhaven, 150–71.

Ashby, E. and Anderson, M. (1981) *The Politics of Clean Air*, Oxford: Clarendon.

Baldock, D., Cox, G., Lowe, P. and Winter, M. (1990) 'Environmentally Sensitive Areas: incrementalism or reform?', *Journal of Rural Studies*, 6: 143–62.

Baldock, D. and Long, T. (1987) *The Mediterranean under Pressure: The Influence of the CAP on Spain and Portugal and the 'IMPs' in France, Greece and Italy*, London: Institute of European Environmental Policy.

Baldock, D. and Lowe, P. (1996) 'The development of European agri-environment policy', in Whitby, M. (ed.) *The European Environment and CAP Reform*, Wallingford, CAB International, 8–25.

Baltic Institute (1994) *Vision and Strategies around the Baltic Sea 2010: Towards a Framework for Spatial Development in the Baltic Sea Region*, Sweden.

Beck, U. (1992) *Risk Society: Towards a New Modernity*, London: Sage. (Translation of the German, *Risikogesellschaft: Auf dem Weges einem andere Moderne*, Frankfurt: Suhrkampf, 1986.)

BirdLife International (1996) *Implementation of the Agri-environment Regulation (EEC 2078/92)*, Sandy: RSPB.

BirdLife International, WWF and World Conservation Union (IUCN) (1995) *Action Plan to 2010: Integrating Agriculture and the Environment*, Sandy: RSPB.

Boardman, R. (1981) *International Organization and the Conservation of Nature*, London: Macmillan.

Bodiguel, M. and Buller, H. (1994) 'Environmental policy and the regions in France', *Regional Policy and Politics*, 4, 3: 92–109.

—— (1996) 'Gestion publique ou gestion privée de l'eau ? Une réflexion franco-britannique', in Mérot, P. and Jogorel, A. (eds) *Hydrologie dans les Pays Celtiques*, Paris: INRA Editions, 443–52.

Boehmer-Christiansen, S. and Skea, J. (1991) *Acid Politics: Environmental and Energy Policies in Britain and Germany*, London: Belhaven.

Boisson, J-M. and Buller, H. (1996) 'The French experience', in Whitby, M.C. (ed.) *The European Environment and CAP Reform*, Wallingford: CAB International, 105–30.

Bowers, J. and Cheshire, P. (1983) *Agriculture, the Countryside and Land Use: An Economic Critique*, London: Methuen.

Bugler, J. (1972) *Polluting Britain: A Report*, London: Penguin.

Buller, H. (1996a) 'Towards sustainable water management: catchment planning in France and Britain', *Land Use Policy*, 13, 4: 289–302.

—— (1996b) 'Privatisation and Europeanisation: the changing context of water supply in Britain and France', *Journal of Environmental Planning and Management*, 39, 4: 461–82.

—— (1996c) *Les administrations nationales et les politiques européennes: comparaison franco-britannique*, paper to the Anniversary Conference of the Ministère de l'Environnement, Paris, 11–12 December.

Buller, H., Lowe, P.D. and Flynn A. (1993) 'National responses to the Europeanisation of environmental policy', in Liefferink, J.D., Lowe, P.D. and Mol, A.P.J. (eds) *European Integration and Environmental Policy*, London: Belhaven, 175–95.

Bulmer, S. (1983), 'Domestic politics and European Community policy making', *Journal of Common Market Studies*, 21: 349–63.

Business Deregulation Task Forces (1994) *Deregulation Task Forces Proposals for Reform*, London, Department of Trade and Industry.

Butson, J. (1993) 'International competitiveness and the "policy edge" in the waste management industry', paper presented to the European Environment Conference, Bristol, 20–21 September.

Butt Philip, A. and Porter, M.H.A. (1993) *Euro-Groups, European Integration and Policy Networks*, Centre for European Industrial Studies, School of Management, University of Bath.

Cabinet Office (1993) *Open Government*, Cmnd 2290, London: HMSO.

Caithness, Earl of (1988) text of speech to RURAL, 18 November, unpublished.

Carter, N. (1992) 'The greening of Labour', in Smith, M. and Spear, J. (eds) *The Changing Labour Party*, London: Routledge, 118–32.

Carter, N. and Lowe, P. (1995) 'The establishment of a cross-sector environment agency', in Gray, T.S. (ed.) *UK Environmental Policy in the 1990s*, London: Macmillan, 38–56.

Caufield, C. (1981) 'Environment: Britain lags, Europe leads', *New Scientist*, 13 August: 416–17.

Christophe, J. (1993) 'The effects of Britons in Brussels: The EC and the culture of Whitehall', *International Journal of Policy and Administration*, 6, 4: 518–37.

Clark, A. (1993) *Diaries*, London: Weidenfeld & Nicolson.

Cole, J. (1995) *As It Seemed to Me: Political Memoirs*, London: Weidenfeld & Nicolson.

Coleman, T. (1992) 'Integrated pollution control: too much to bear and too much to bare?', keynote address to the BICS Conference, 9 December.

Collins, K. and Earnshaw, D. (1992) 'The implementation and enforcement of European Community legislation', *Environmental Politics*, 1, 4: 213–49.

Commission of the European Communities (CEC) (1990a) 'Council Regulation 1210/90 on the establishment of the European Environment Agency and the European environment information and observation network', OJ L120, 11 May.

—— (1990b) 'Lomé IV Convention', in the *ACP-EEC Courier*, 120, March–April.

—— (1990c) 'Green Paper on the urban environment', COM(90)218.

—— (1991a) CORINE *Biotopes Manual*, CEC DGXI-EEA-TF, EUR 13231.

—— (1991b) 'Results of the CORINE Programme'. SEC(91)958 final, 28 May.

—— (1991c) *Europe 2000: Outlook for Development of the Community's Territory*, Brussels: CEC.

—— (1992a) 'Council Regulation 2078/92', OJ L215, 30 July.

—— (1992b) *Towards Sustainability: A European Community Programme of Policy and Action in Relation to the Environment and Sustainable Development*, COM(92)23 final.

—— (1993a) 'An open and structured dialogue between the Commission and special interest groups', OJ C 63/02, 5 March.

—— (1993b) 'Proposal for a Council Directive on Integrated Pollution Prevention and Control', COM(93)423 final, 14 September.

—— (1993c) *Growth, Competitiveness, Employment: The Challenge and Ways Forward into the 21st Century*, COM(93)700.

—— (1995a) 'Amended proposal for a Council Directive on Integrated Pollution Prevention and Control', COM(95)85 final, June.

—— (1995b) European Commission Report on Waste Management Policy, COM(95)522 final, 8 November.

—— (1995c) *Europe 2000+, CEC DGXVI.*

—— (1995d) 'Proposal for a Council Decision on a Community Action Programme promoting non-governmental organisations active in the field of environmental protection', COM(95)573 final, 8 December.

—— (1995e) 'Study on alternative strategies for the development of relations in the field of agriculture between the EU and the associated countries with a view to future accession of these countries', Agricultural Strategy Paper, communication by Commissioner Fischler, December 1995.

—— (1995f) 'The Common Transport Policy Action Programme 1995–2000', COM(95)302.

—— (1996a) 'Proposal for a European Parliament and Council decision on the review of the European Community Programme of policy and action in relation to the environment and sustainable development Towards Sustainability', COM(95)647 final, 24 January.

—— (1996b) 'Call for the submission of proposals for the promotion of representative European organisations working in the field of the environment', OJ C 71/12, 9 March.

—— (1996c) 'List of non-governmental organisations that have received Community funding', OJ C 114/02, 19 April.

—— (1996d) *RTD Info*, February, CEC DGXII.

—— (1996e) 'European Community Water Policy', COM(96)59 final, 21 February.

—— (1996f) *The Cork Declaration: A Living Countryside*, Resolution issued by the Cork conference, November.

Constitution Unit (1996) *Regional Government in England: The Independent Inquiry into the Implementation of Constitutional Reform*, University College London and Joseph Rowntree Foundation, London: The Constitution Unit.

Council of Europe (1995a) *Biodiversity Strategy for Europe*, Strasbourg: Council of Europe.

—— (1995b) *Environmental Action Plan for Central and Eastern Europe*, Strasbourg: Council of Europe.

Cox, G., Lowe, P. and Winter, M. (1990) *The Voluntary Principle in Conservation*, Chichester: Packard.

Crockers (1993) *How Recent Developments in Environmental Law May Affect Local Authorities – with Particular Reference to the Environmental Protection Act (1990) and European Directives*, London: Crockers Solicitors.

Dalton, R.J. (1992) 'Alliance patterns of the European environmental movement', in Rudig, W., *Green Politics Two*, Edinburgh: Edinburgh University Press, 37–58.

David, H., Klatte, E. and Van Ermen, R. (1994) 'A long way', in *EEB Twentieth Anniversary*, Brussels: European Environmental Bureau.

Davies, J.C. and Davies, B. (1975) *The Politics of Pollution*, Indianapolis: Pegasus, 2nd edn.

Denman, R. (1995) 'Missed chances: Britain and Europe in the twentieth century', *Political Quarterly*, 66, 1: 36–47.

Department of the Environment (DoE) (1975) *Report of the Coastal Pollution Research Committee of the Water Pollution Research Laboratory*, London: HMSO.

—— (1976) *Pollution Control in Britain: How It Works*, London: HMSO.

—— (1978) *Pollution Control in Great Britain: How It Works* (2nd edn), Pollution Paper 9, London: HMSO.

—— (1980) *Advice Note of November 1980: Implementation in England and Wales of the EC Directive on the Quality Required of Shellfish Waters*, London: DoE.

—— (1986) *Waste Disposal Law: Amendment*, London: DoE.

—— (1993a) *Integrated Pollution Control: A Practical Guide*, London: HMSO.

—— (1993b) *The Links between the Department of the Environment and the Institutions of the European Community: A Review*, December, London: HMSO, Annex 1.

—— (1994) *Towards Sustainability: Government Action in the UK – An Interim Progress Report*, London: HMSO.

—— (1995a) *EC Integrated Pollution Prevention and Control Directive: Written Correspondence to Consultees*, 19 September.

—— (1995b) *How It Will Apply in Great Britain: Habitats Directive*, London: DoE.

—— (1995c) *Letter to Dr Caroline Jackson MEP*, copied to all UK MEPs, 11 October.

—— (1995d) news release, 31 March.

Department of the Environment/Welsh Office (1988) *Waste Disposal Law Amendment: Decisions Following Public Consultation*, London: DoE/Welsh Office.

—— (1989) *The Role and Functions of Waste Disposal Authorities: Announcement of Government Decisions*, London: DoE/Welsh Office.

—— (1993) *Water Charges: The Quality Framework*, London: DoE/Welsh Office.

—— (1995) *Making Waste Work: A Strategy for Sustainable Waste Management in England and Wales*, CM 3040, London: HMSO.

Environmental Data Services (ENDS) (1991) 'Playing the Euro-Lottery', *Environmental Data Services*, report 201, October: 2.

—— (1995a) 'Brussels trims plans for legislation in 1996', *Environmental Data Services*, report 250, November: 35–6.

—— (1995b) 'Environment Agency to take lead role in sustainable waste management', *Environmental Data Services*, report 251, December: 15.

—— (1995c) 'IPPC Directive to force major changes in IPC', *Environmental Data Services*, report 245, June, 43–4.

—— (1996a) 'Brussels bids to revive flagging environment programme', *Environmental Data Services*, report 253, February: 24–36.

—— (1996b) 'Industry, HMIP Guide on Environmental Analysis for IPC', *Environmental Data Services*, report 255, April: 34.

—— (1996c) 'IPC bites in the electricity sector – but BATNEEC takes a back seat' *Environmental Data Services*, report 254, March: 18–21.

Environment Watch Western Europe (EWWE) (1995) 'Waste strategy review is top priority, EU's Enthoven says', press release, 6 January, 7.

European Community Task Force (1989) *Environment and the Internal Market*, Brussels: European Commission.

European Environment Agency (EEA) (1994) *European Environment Agency: Putting Information to Work*, Copenhagen: EEA.

European Information Service (EIS) (1996), 'Waste strategy under review', *European Information Service*, 167 (February): 25–6.

Evans, D. (1973) *Britain in the EEC*, London: Victor Gollancz.

Flynn, A. and Lowe, P. (1992) 'Greening the Tories: the Conservative Party and the environment', in Rudig, W. (ed.), *Green Politics Two*, Edinburgh: Edinburgh University Press, 9–35.

Frankel, M. (1974) *The Alkali Inspectorate: The Control of Air Pollution*, London: Social Audit.

Friends of the Earth (FoE) (1990) *How Green Is Britain? The Government's Environmental Record*, London: Hutchinson Radius.

Gallagher, E. (1996) text of speech at first public briefing on the Environment Agency, 'Framework for the Future', 6 February at the Queen Elizabeth II Conference Centre, London.

George, S. and Haythorne, D. (1996) 'The British Labour Party', in Gaffney, J. (ed.) *Political Parties and the European Union*, London: Routledge, 110–20.

George, S. (1990) *An Awkward Partner: Britain in the European Community*, Oxford: Oxford University Press.

George, S. (ed) (1992) *Britain and the European Community: The Politics of Semi-Detachment*, Oxford: Oxford University Press.

Gibbons, M., Limoges, C., Nowotny, H. Schwarzman, S., Scott, P. and Trow, M. (1994) *The New Production of Knowledge: The Dynamics of Science and Research in Contemporary Societies*, London: Sage.

Giddens, A. (1990) *The Consequences of Modernity*, Cambridge: Polity.

—— (1991) *Modernity and Self-Identity: Self and Society in the Late Modern Age*, Cambridge: Polity.

Golub, J (1996) 'Sovereignty and subsidiarity in EU environmental policy', *Political Studies*, 44: 686–703.

Grant, W. (1993) 'Pressure groups and the European Community: an overview', in Mazey, S. and Richardson, J. (eds) *Lobbying in the European Community*, Oxford: Oxford University Press, 27–46.

Greenwood, J., Grote, J. and Ronit, K. (eds) (1992) *Organised Interests and the European Community*, London: Sage.

Grimmett, R. and Jones, T.A. (1989) *Important Bird Areas in Europe*, Cambridge: BirdLife International.

Gummer, J. (1994) *Europe, What Next? Environment, Policy and the Community*, a speech to the ERM Environment Forum organised by the Green Alliance hosted by the Royal Society of Arts, 20 December, Green Alliance, 2.

Haigh, N. (1984) *EEC Environmental Policy & Britain : An Essay and a Handbook*, London: Environmental Data Services.

—— (1986) 'Devolved responsibility and centralisation: the effects of EEC environmental policy', *Public Administration*, 64: 197–207.

—— (1987) *EEC Environmental Policy and Britain; An Essay and a Handbook*, 2nd edn, Harlow: Longman.

—— (1989), *EEC Environmental Policy and Britain*, 2nd rev. edn. Harlow: Longman.

—— (1992) *Manual of Environmental Policy*, Harlow: Longman.

—— (1994a) *Manual of Environmental Policy: The EC and Britain*, London: Longman.

—— (1994b) 'The introduction of the precautionary principle in the UK', in O'Riordan, T. and Cameron, J. (eds) *Interpreting the Precautionary Principle*, London: Earthscan.

—— (1995a) *Environmental Protection in the DoE (1970–1995) or One and A Half Cheers for Bureaucracy*, paper delivered at a Conference to mark the 25th Anniversary of the DoE, London, 30 October.

—— (1995b) (ed.) *Manual of Environmental Policy*, London: Institute for European Environmental Policy.

—— (1996) 'Climate change policies and politics in the European Community', in O'Riordan, T. and Jäger, J. (eds) *Politics of Climate Change*, London: Routledge, 155–85.

Haigh, N. and Lanigan, C. (1995) 'The impact of the EU on UK environmental policy making', in Gray, T. (ed.) *UK Environmental Policy in the 1990s*, Basingstoke: Macmillan, 18–37.

Haigh, N. *et al.* (1986) *European Community Environmental Policy in Practice*, vol. 1, *Comparative Report: Waste and Water in Four Countries – A Study of the Implementation of the EEC Directives in France, Germany, Netherlands and the UK*, London: Graham & Trotman.

Hajer, M. (1995) *The Politics of Environmental Discourse*, Oxford: Oxford University Press.

Hall, P. (1993) 'Policy paradigms, social learning and the state', *Comparative Politics*, 25, 3: 275–96.

Hallo, R. (1995) *Greening the Treaty II: Sustainable Development in a Democratic Union – Proposals for the 1996 Intergovernmental Conference*, Utrecht: Stichting Natuur en Milieu.

Harris, D. (1974) 'The law relating to the pollution of inland waters: international aspects', in McKnight, A.D., Marstrand, P.K. and Sinclair, T.C. (eds) *Environmental Pollution Control: Technical, Economic and Legal Aspects*, London: Allen & Unwin, 77–88.

Hawkins, R. (1984) *Environment and Enforcement*, Oxford: Clarendon.

Health and Safety Executive (1986) *Health and Safety: Industrial Air Pollution 1985*, London: HMSO.

Heidenheimer, A.J., Heclo, H. and Teich Adams, C. (1990) *Comparative Public Policy: The Politics of Social Choice in America, Europe and Japan*, London: Macmillan.

Héritier, A. (1992) 'Policy-Netzwerkanalyse als Untersuchungsinstrument im europäischen Kontext', *Politisches Vierteljahresschrift* 34, 24: 432–47.

Héritier, A. *et al.* (1994) *Die Voränderung von Staatlichkeit in Europa: Ein regulativer Wettbewerb: Deutschland, Grossbritannien, Frankreich*, Opladen: Leske & Budrich.

Her Majesty's Inspectorate of Pollution (HMIP) (1990) *Forward Look 1990–91 to 1994–95*, London: HMIP.

—— (1992) *Chief Inspector's Guidance Notes – Process Guidance Note IPR5/1 – Merchant & In House Chemical Waste Incineration*, London: HMSO.

—— (1993) *Chief Inspector's Guidance Note – Process Guidance Note IPR4/5 – Batch Manufacture of Organic Chemicals in Multipurpose Plant*, London: HMSO.

—— (1995) *Chief Inspector's Guidance Notes – S2 1.03 – Combustion Processes: Compression Ignition Engines 50 MW(th) and Over*, London: HMSO.

Heseltine, M. (1991) 'Text of speech to an all-party Parliamentary back-bench Committee' given in spring 1991, unpublished.

Hey, C. and Brendle, U. (1992) *Environmental Organisations and the EC: Action Options of Environmental Organisations for Improving Environmental Consciousness and Environmental Policy in the European Community*, Freiburg: Institut für regionale Studien in Europa e.V. (EURES).

—— and Brendle, U. (1994) *Towards a New Renaissance: A New Development Model*, Brussels: European Environmental Bureau.

Hildebrand, P. (1992) 'The European Community's environmental policy, 1957 to 1992: from incidental measures to an international regime', *Environmental Politics*, 1, 4: 13–44.

Hill, M. *et al.* (1989) 'Non decision making in pollution control in Britain: nitrate pollution, the EEC Drinking Water Directive and agriculture', *Policy and Politics*, 17: 227–40.

House of Commons Select Committee on the Environment (1990) Fourth Report 1989–90: *Pollution of Beaches* London: HMSO.

—— (1995) Third Report 1994–95: *Pollution in Eastern Europe*, London: HMSO.

House of Lords Select Committee on the European Communities (1989) Fifteenth Report 1988–89: *Habitats and Species Protection*, London: HMSO.

—— (1991) Tenth Report 1990–91: *Municipal Waste Water Treatment*, London: HMSO.

—— (1995) Fifth Report 1994–95: *European Environment Agency*, London: HMSO.

—— (1995) First Report 1994–95: *Bathing Water*, London: HMSO.

—— (1995) Seventh Report 1994–95: *Bathing Water Revisited*, London: HMSO.

—— (1996) Fourth Report 1995–96: *Drinking Water*, London: HMSO.

House of Lords Select Committee on Sustainable Development (1996) Session 1995–96 *Report from the Select Committee Vol I*, London: HMSO.

Hughes, D. (1995) 'The status of the precautionary principle in law', *Journal of Environmental Law*, 7: 224–44.

Hull, R. (1993) 'Lobbying Brussels: a view from within', in Mazey, S. and Richardson, J.J. *Lobbying in the European Community*, Oxford: OUP, 82–94.

Humphreys, J. (1995) *A Way through the Woods*, London: Department of the Environment.

Huxley, J.S. (1947) *Conservation of Nature in England and Wales*, Cmnd 7122, London: HMSO.

Ireland, F. (1967) *Chief Inspector's 104th Annual Report for the Alkali Inspectorate*, London: HMSO.

Jachtenfuchs, M. (1991) 'The European Community and the protection of the ozone layer', *Journal of Common Market Studies*, 29: 261–77.

Jachtenfuchs, M. and Huber, M. (1993) 'Institutional learning in the European Community: the response to the greenhouse effect', in Liefferink, J.D., Lowe, P.D. and Mol, A.P.J. (eds) *European Integration and Environmental Policy*, London: Belhaven, 36–58.

Jans, J.H. (1990) 'Legal grounds of European environmental policy', in *Milieu* (Netherlands Journal of Environmental Sciences) *Environmental Policy in the EC*, vol. 5, 1990/6, Meppel: Boom Publications.

Jasanoff, S. (1996) *Harmonization and the Politics of Reasoning Together*, unpublished working paper.

Jenkins, S. (1995) *Accountable to None: The Tory Nationalisation of Britain*, London: Hamish Hamilton.

Jimenez-Beltran, D. (1994) *European Environment Agency: Putting Information to Work*, Copenhagen: EEA.

John, P. (1994) 'Room with a view', *Local Government Chronicle*, 7 July, 10–11.

Johnson, B. (1971) 'Common Market v. Environment', *The Ecologist*, 1, 11: 10–14.

Johnson, S.P. (1979) *The Environment Policy of the European Communities*, London: Graham & Trotman.

Johnson, S.P. and Corcelle, G. (1989) *The Pollution Control Policy of the European Communities*, London: Graham & Trotman.

Jordan, A. (1993) 'IPC and the evolving style and structure of environmental regulation in the UK', *Environmental Politics*, 2, 3: 405–27.

Jordan, A., Ward, N. and Buller, H. (1996) 'Sun, sea, sand and pollution: the implementation of the EU Bathing Waters Directive in Britain and France', paper to the European Consortium for Political Research Joint Sessions, Oslo, April.

Jordan, G. and Richardson, J. (1983) 'Policy communities: the British style and European style?', *Policy Studies Journal*, 11: 603–11.

Judge, D. and Earnshaw, D. (1993) 'Weak European Parliament influence? A study of the Environment Committee of the European Parliament', *Government and Opposition*, 28: 262–76.

Kinnersley, D. (1994) *Coming Clean: The Politics of Water and the Environment*, London: Penguin.

Koppen, I.J. (1988) *The European Community's Environmental Policy: From the Summit in Paris, 1972, to the Single European Act, 1987*, EUI working paper 88/328, Florence: European University Institute.

——— (1993) 'The role of the European Court of Justice', in Liefferink, J.D., Lowe, P.D. and Mol, A.P.J. (eds), *European Integration and Environmental Policy*, London: Belhaven.

Kousis, M. (1994) 'Environment and the State in the EU periphery: the case of Greece', in Baker, S. *et al.* (eds), *Protecting the Periphery: Environmental Policy in the Peripheral Regions of the EU*, Ilford: Frank Cass, 118–35.

Krämer, L. (1987) 'The Single European Act and environment protection: reflections on several new provisions in Community law', *Common Market Law Review*, 24: 659–88.

——— (1989) 'The open society, its lawyers and its environment', *Journal of Environmental Law*, 1: 1–9.

——— (1991) 'Participation of environmental organisations in the activities of the EEC', in M. Fuhr *et al.* (eds), *Participation Rights in European Perspective*, New York and Frankfurt am Main.

——— (1995) *EC Treaty and Environmental Law* (2nd edn), London: Sweet & Maxwell.

Ladeur, K.H. (1996) *The New European Agencies: The European Environmental Agency and Prospects for a European Network of Environmental Administrations*, EUI working paper RSC 95/50, Florence: European University Institute.

Laffin, M. (1986) *Professionalism and Policy: The Role of Professions in Central–Local Government Relationships*, Aldershot: Gower.

Lash, S., Szerszynski, B. and Wynne, B. (1996) *Risk, Environment and Modernity: Towards a New Ecology*, London: Sage.

La Spina, A. and Sciortino, G. (1993) 'Common agenda, southern rules: European integration and environmental change in the Mediterranean states', in Liefferink, J.D., Lowe, P.D. and Mol, A.J.P. (eds), *European Integration and Environmental Policy*, London: Belhaven.

Latour, B. (1995) 'The "pedofil" of Boa Vista: a photo-philosophical montage', trans. B. Simon and K. Verresen, *Common Knowledge*, 4: 1.

Laurence, J. and Wynne, B. (1989) 'Transporting waste in the European Community: a free market?', *Environment*, 31, 6: 12–17.

Lee, N. (1992) 'Environmental policy', in Bulmer, S. (ed), *Britain and the European Community: Membership Evaluated*, London: Pinter.

Lequesne, C. (1993) *Paris-Bruxelles: comment se fait la politique Européenne de la France*, Paris: Presses de la Fondation Nationale des Sciences Politiques.

Levitt, R. (1980) *Implementing Public Policy*, London: Croom Helm.

Liddell, P. (1974) 'The new water industry', *Municipal and Public Services Journal*, 82: 249–55.

Liefferink, J.D., Lowe, P.D. and Mol, A.P.J. (eds) (1993) *European Integration and Environmental Policy*, London: Belhaven.

Liefferink, J.D. and Mol, A.J. (1993) 'Environmental policy making in the European Community: an evaluation of theoretical perspectives', in Liefferink, J.D., Lowe, P.D. and Mol, A.P.J. (eds), *European Integration and Environmental Policy*, London: Belhaven.

Local Government Management Board (LGMB) (1993) *Towards Sustainability: The EC's Fifth Action Programme on the Environment – A Guide for Local Authorities*, Luton: LGMB.

——— (1996) *Review of the EU's Fifth Environmental Action Programme*, Luton: LGMB.

Long, T. (1995) 'Shaping public policy in the European Union: a case study of the Structural Funds', *Journal of European Public Policy*, 2: 4, 672–80.

Lorrain, D. (1991) 'Public goods and private operators in France', in Batley, R. and Stoker, G. (eds) *Local Government in Europe*, London: Macmillan, 89–109.

Lowe, P. and Bodiguel, M. (eds) (1990) *Rural Studies in Britain and France*, London: Belhaven.

Lowe, P., Cox, G., MacEwan, M., O'Riordan, T. and Winter, M. (1986) *Countryside Conflicts: The Politics of Farming, Forestry, and Conservation*, Aldershot: Gower.

Lowe, P. and Flynn, A. (1989) 'Environmental politics and policy in the 1980s', in Mohan, J. (ed.) *The Political Geography of Contemporary Britain*, London: Macmillan, 255–79.

Lowe, P. and Goyder, J. (1983) *Environmental Groups in Politics*, London: Allen & Unwin.

Lowe, P. and Murdoch, J. (1994) 'European review 1992/3' in Gilg, A. (ed.) *Progress in Rural Policy and Planning*, vol. 4, London: Wiley.

—— (1995), 'European review 1993/4', in Gilg, A. (ed.), *Progress in Rural Policy and Planning*, vol. 5, London: Wiley.

Ludlow, P. (1991) 'The European Commission', in Keohane, R.O. and Hoffmann, S. (eds) *The New European Community*, Boulder, Col.: Westview.

McAuslan, P. (1991) 'The role of the courts and other judicial type bodies in environmental management', *Journal of Environmental Law*, 3: 195–208.

McCormick, J. (1991) *British Politics and the Environment*, London: Earthscan.

—— (1995) *The Global Environmental Movement*, Chichester: Wiley.

McCracken, D.I. and Bignal, E. (1995) *Farming on the Edge: The Nature of Traditional Farmland in Europe*, Proceedings of the 4th European Forum on Nature Conservation and Pastoralism, Peterborough: Joint Nature Conservation Committee.

McLaughlin, A.N. and Greenwood, J. (1995) 'The management of interest representation in the European Union', *Journal of Common Market Studies*, 33: 143–56.

McLaughlin, A.N., Jordan, A.G. and Maloney, W.A. (1993) 'Corporate lobbying in the European Community', *Journal of Common Market Studies*, 31: 191–212.

Macrory, R. (1987) 'The United Kingdom', in Enyedi, G., Giswijt, J. and Rhode, B. (eds) *Environmental Policies in East and West*, London: Taylor & Graham.

—— (1991) 'Environmental law: shifting discretions and the new formalism', in Lomas, O. (ed.) *Frontiers of Environmental Law*, London: Chancery Law.

Macrory, R. and Hession, M. (1996) 'The European Community and climate change: the role of law and legal competence', in O'Riordan, T. and Jäger, J. (eds), *Politics of Climate Change*, London: Routledge, 106–54.

Madel, E. and Wynne, B. (1987) 'Decentralised regulation and technical discretion: the UK', in Wynne, B. (ed.), *Risk Management and Hazardous Waste: Implementation and the Dialectics of Credibility*, London/Berlin/New York: Springer Verlag, 195–224.

Mahler, E.A.J. (1967) 'Standards of emission under the Alkali Act', paper presented to the International Clean Air Congress, London, October, 1966. Reproduced in *103rd Annual Report of the Alkali Inspectorate*, London: HMSO.

Maloney, W. and Richardson, J.J. (1994) 'Water policy making in England and Wales: policy communities under pressure', *Environmental Politics*, 3, 4: 110–37.

Maloney, W. and Richardson, J.J. (1995) *Managing Policy Change in Britain: The Politics of Water*, Edinburgh: Edinburgh University Press.

Marks, G. (1992) 'Structural policy in the European Community', in Sbragia, A. (ed.) *Euro-Politics*, Washington, DC: Brookings Institution.

Matthews, D. and Pickering, J. (1996) *The Role of the Firm in the Evolution of European Environmental Rules: The Case of the Water Industry and the European Drinking Water Directive*, discussion paper 92, London: National Institute of Economic and Social Research.

Mawson, J. (1995) *The Operation of the Single Regeneration Budget*, Birmingham: University of Birmingham.

Mawson, J. and Spencer, K. (1997) 'The Government Offices for the English regions: towards regional governance', *Policy and Politics* 25, 71–84.

Mazey, S. and Richardson, J.J. (1992a), 'British pressure groups and the EC: the challenge of Brussels', *Parliamentary Affairs*, 45, 1: 92–108.

—— (1992b) 'Environmental groups and the EC: challenges and opportunities', *Environmental Politics*, 1, 4: 110–28.

—— (1994) 'Policy co-ordination in Brussels: environmental and regional policy', in Baker, S. *et al.* (eds) *Protecting the Periphery: Environmental Policy in the Peripheral Regions of the EU*, Ilford: Frank Cass, 22–44.

Melucci, A. (1989) *Nomads of the Present: Social Movements and Individual Needs in Contemporary Society*, London: Hutchinson Radius.

Merrill, P. (1994) 'MGM (My God More!) Productions presents … Chemical Tranche 1 – A Documentary Drama', *HMIP Bulletin*, 30, 16–17.

Ministry of Agriculture, Fisheries and Food, CAP Review Group (1995) *European Agriculture: The Case for Radical Reform*, London: MAFF.

Moore, N. W. (1987) *The Bird of Time: The Science and Politics of Nature Conservation*, Cambridge: Cambridge University Press.

Morphet, J. (1993) *A Guide to the EU's Fifth Environmental Action Programme 'Towards Sustainability'*, Luton: Local Government Management Board.

—— (1994) 'The Committee of the Regions', *Local Government Policy Making*, 20, 5: 56–60.

—— (1995) 'Greening development control ', in Oc, T. and Trench, S. (eds), *Current Issues in Planning*, vol. 2, Aldershot: Avebury.

Morphet, J. and Hams, T. (1994) 'Responding to Rio: a local authority approach', *Journal of Environmental Management*, 37, 4: 477–82.

Morphet, J., Hams, T., Taylor, D., Lusser, H., Jacobs, M. and Levett, R. (1994) *Greening Your Local Authority*, London: Longman.

Muller, P. (1994) 'Europe, de la Communauté à l'Union', *Pouvoirs*, 69: 63–75.

National Audit Office (1991) *Control and Monitoring of Pollution: Review of the Pollution Inspectorate*, London: HMSO.

National Rivers Authority (NRA) (1991) *Bathing Water Quality in England and Wales – 1990*, Water Quality Series no. 3, Bristol: NRA.

—— (1994) 'Evidence to the Committee', in House of Lords Select Committee on the European Communities, *Bathing Water*, HL paper 6–I, London: HMSO, 14–31.

—— (1995) *Bathing Water Quality in England and Wales – 1994*, Water Quality Series no. 22, Bristol: NRA.

Nature Conservancy Council (NCC) (1984) *Nature Conservation in Great Britain*, Peterborough: NCC.

Nicholson, E.M. (1970) *The Environmental Revolution: A Guide for the New Masters of the World*, London: Hodder & Stoughton.

Nicholson, E.M. (1987) *The New Environmental Age*, Cambridge: Cambridge University Press.

Organisation for Economic Co-operation and Development (1994) *Environmental Performance Reviews: United Kingdom*, Paris: OECD.

O'Riordan, T. and Jordan, A. (1995) 'The precautionary principle in contemporary environmental politics', *Environmental Values*, 4, 3: 191–212.

O'Riordan, T. and Weale, A. (1989) 'Administrative reorganisation and policy change: the case of Her Majesty's Inspectorate of Pollution', *Public Administration*, 67, autumn: 277–94.

Osborn, D. (1992) 'The impact of EC environmental policies on UK public administration', *Environmental Policy and Practice*, 2: 199–209.

Pellegrom, S. (1996) 'The constraints of daily work in Brussels: how relevant is the input from the national capitals?', in Liefferink, J.D. and Andersen, M.S. (eds) *The Innovation of EU Environmental Policy*, Copenhagen: Scandinavian University Press.

Pestellini, F. (1992) 'Free trade of hazardous wastes? Problems and prospects of the implementation of the Basel Convention within the European Community', in *European Environment*, 2, 5: 5–10.

Peterken, G. (1996) *Ancient Woodland*, Cambridge: Cambridge University Press.

Peters, B.G. (1994) 'Agenda setting in the European Community', *Journal of European Public Policy*, 1, 1: 9–26.

Porter, M.H.A. (1995) *Interest Groups, Advocacy Coalitions and the EC Environmental Policy Process: A Policy Network Analysis of the Packaging and Packaging Waste Directive*, unpublished DPhil registered at the University of Bath.

—— (1996) *The Development of the Urban Waste Water Treatment Directive*, draft working paper for the CIBR, University of Bath.

Potter, C. and Lobley, M. (1990) 'Adapting to Europe: conservation groups and the EC', *Ecos*, 11, 3: 3–7.

Prate, A. (1995) *La France en Europe*, Paris: Economica.

Princen, T. and Finger, M. (1994) *Environmental NGOs in World Politics*, London: Routledge.

Ratcliffe, D. (ed.) (1977) *A Nature Conservation Review*, Nature Conservancy Council, Cambridge: Cambridge University Press.

Renshaw, D. (1980) 'Water quality objectives', in Gower, A. (ed.), *Water Quality in Catchment Ecosystems*, Chichester: Wiley.

Rhodes R.A.W. (1986) *European Policy-Making, Implementation and Subcentral Governments: A Survey*, Maastricht: European Institute of Public Administration.

Richardson, G., Ogus, P. and Burrows, P. (1983) *Policing Pollution*, Oxford: Oxford University Press.

Richardson, J.J. (ed.) (1982) *Policy Styles in Western Europe*, London: Allen & Unwin.

Richardson, J.J. and Jordan, A.G. (1979), *Governing Under Pressure: The Policy Process in a Post Parliamentary Democracy*, Oxford: Blackwell.

—— (1987) *British Politics and the Policy Process*, London: Unwin Hyman.

Richardson, J.J. and Watts, N.S.J. (1985), *National Policy Styles and the Environment: Britain and West Germany Compared*, Berlin: Internationales Institut für Umwelt und Gesellschaft.

Robinson, M. (1992) *The Greening of British Party Politics*, Manchester: Manchester University Press.

Rose, C. (1990) *The Dirty Man of Europe: The Great British Pollution Scandal*, London: Simon & Schuster.

Rousseau, A. (1993) *Vers une modification du paramètre pesticide de la directive eau potable*, Brussels: Bureau Européen de l'Environnement.

Royal Commission on Environmental Pollution (RCEP) (1972) *Three Issues in Industrial Pollution*, Second Report, Cmnd 4894, London, HMSO.

—— (1976) *Air Pollution: An Integrated Approach*, Fifth Report, Cmnd. 6371, London: HMSO.

—— (1984) *Tackling Pollution: Experience and Prospects*, Tenth Report, Cmnd. 9149, London: HMSO.

—— (1985) *Managing Waste: The Duty of Care*, Eleventh Report, Cmnd 9675, London: HMSO.

—— (1988) *Best Practicable Environmental Option*, Twelfth Report, Cmnd 310, London, HMSO.

Royal Society for the Protection of Birds, BirdLife International, WWT, Game Conservancy Trust, British Trust for Ornithology, Hawk and Owl Trust, Wildlife Trusts, and the National Trust (1996) *Birds of Conservation Concern in the UK, Channel Islands and Isle of Man*, Sandy: RSPB.

Rucht, D. (1993) ' "Think globally, act locally"? Needs, forms and problems of cross-national co-operation among environmental groups', in Liefferink, J.D., Lowe, P.D. and Mol, A.P.J. (eds), *European Integration and Environmental Policy*, London: Belhaven, 75–95.

Sands, P. (1990) 'European Community environmental law: legislation, European Court of Justice and common interest groups', *Modern Law Review*, 53, 5: 685–98.

Sbraiga, A. (1996) 'Environmental policy', in Wallace, H. and Wallace, W. (eds) *Policy Making in the European Union*, Oxford: Oxford University Press.

Schendelen, M.P.C.M. van (ed.) (1993) *National Public and Private EC Lobbying*, Dartmouth: Aldershot.

Schmidheiny, S. (1992) *Changing Course: A Global Business Perspective on Development and the Environment*, Cambridge, Mass.: Massachusetts Institute of Technology.

Smith, A. (1996) 'Voluntary schemes and the need for statutory legislation: the case of integrated pollution control', *Business Strategy and the Environment*, 5, 1, 81–7.

Stanners, D. and Bordeau, P. (eds) (1995) *1995 Europe's Environment: The Dobris Assessment*, Copenhagen: European Environment Agency.

Straw, J. (1995) *A Voice for England?*, London: Labour Party.

Taylor, D., Diprose, G. and Diffy, M. (1986) 'EC environmental policy and the control of water pollution: the implementation of Directive 76/464 in perspective', *Journal of Common Market Studies*, 24: 225–46.

Tindale, S. (1992) 'Learning to love the market: Labour and the European Community', *Political Quarterly* 63, 2: 276–300.

Tinker, J. (1972) 'Britain's environment: Nanny knows best', *New Scientist*, 53, 786 (9 March): 530.

Tucker, G.M. and Heath, M. (1994) *Birds in Europe: Their Conservation Status*, Cambridge: BirdLife International.

Tunnicliffe, M.F. (1975) 'The United Kingdom approach and its application by central government: standards of emission for scheduled processes', paper presented to the International Clean Air and Pollution Control Conference, 1975, reproduced in Health and Safety Executive, *Industrial Air Pollution Annual Report*, 1975, London: HMSO.

UK Government (1990) *This Common Inheritance, Britain's Environmental Strategy*, Cmnd 1200, London: HMSO.

—— (1994a) *Biodiversity: The UK Action Plan*, Cmnd 2428, London: HMSO.

—— (1994b) *Sustainable Development: The UK Strategy*, Cmnd 2426, London: HMSO.

—— (1996) *A Partnership of Nations*, London: HMSO.

Van Ermen, R. (1991) *Relations among Environmental NGOs and Networks at the EEC Level*, Brussels: European Environmental Bureau.

Vogel, D. (1983) 'Comparing policy styles: environmental protection in the US and Britain', *Public Administration Bulletin*, 42: 65–78.

—— (1986) *National Styles of Regulation: Environmental Policy in Great Britain and the United States*, Ithaca, NY: Cornell University Press.

—— (1993) 'The making of EC environmental policy', in Andersen, S.S. and Eliassen, K.A. (eds), *Making Policy in Europe: The Europeification of National Policy-making*, London: Sage.

Vonkeman, G. (1994) 'After 20 years: a reflection', in *EEB Twentieth Anniversary*, Brussels: European Environmental Bureau.

Von Weizsäcker, E.U. (1995) *Earth Politics*, London and Jersey: Zed Books.

Waldegrave, W. (1985) 'The British approach', *Environmental Policy and Law*, 15, 3–4: 106–15.

Wallace, H. (1971) 'The impact of the European Communities on national policy making', *Government and Opposition* 6, 4: 520–38.

—— (1972) 'The impact of the European Communities on national policy-making', in Hodges, M. (ed.), *European Integration*, Harmondsworth: Penguin, 285–303.

—— (1995) 'Britain out on a limb?', *Political Quarterly*, 66, 1: 46–58.

Ward, H. (1993) 'Purity and danger: The politicisation of drinking water quality in the eighties', in Mills, M. (ed.), *Prevention, Health and British Politics*, Aldershot: Avebury, 123–39.

Ward, N. (1996a) 'Pesticides, pollution and sustainability', in Allanson, P. and Whitby, M. (eds), *The Rural Economy and the British Countryside*, London: Earthscan, 40–61.

—— (1996b) 'Surfers, sewage and the new politics of pollution', *Area*, 28: 331–8.

Ward, N., Buller, H. and Lowe, P. (1995) *Implementing European Environmental Policy at the Local Level: The UK Experience With Water Quality Directives*, vols I and II, Newcastle upon Tyne: Centre for Rural Economy Research Report, University of Newcastle upon Tyne.

—— (1996), 'The Europeanisation of local environmental politics: bathing water pollution in south-west England', *Local Environment*, 1, 1: 21–32.

Ward, S.J. (1995) 'The politics of mutual attraction: UK local authorities and the Europeanisation of environmental policy', in Gray, T. (ed.), *UK Environmental Policy in the 1990s*, Basingstoke: Macmillan, 101–21.

Ward, S.J. and Lowe, P.D. (1994) *Adaptation, Participation and Reaction: British Local Government – EU Environmental Relations*, Centre for Rural Economy Research Report, University of Newcastle upon Tyne

Ward, S.J., Talbot, H. and Lowe, P.D. (1995) 'Environmental agencies and Europe: a case of missed opportunities', *Ecos*, 16, 2: 47–53.

Ward, S.J. and Williams, R. (1997) 'From hierarchy to networks? Sub-central government and EU urban environmental policy', to appear in *Journal of Common Market Studies*, 35, 3 (forthcoming).

Waterton, C., Grove-White, R., Rodwell, J. and Wynne, B. (1995) *CORINE: Databases and Nature Conservation – the New Politics of Information in the European Union*, report to WWF-UK, Lancaster University: Centre for the Study of Environmental Change.

Waterton, C. and Wynne, B. (1996) 'Building the European Union: science and the cultural dimensions of environmental policy', *Journal of European Public Policy*, 3, 3: 421–40.

Weait, M. (1989) 'The letter of the law? An enquiry into reasoning and formal enforcement in the Industrial Air Pollution Inspectorate', *British Journal of Criminology*, 29, 1: 57–70.

Weale, A. (1992) *The New Politics of Pollution*, Manchester: Manchester University Press.

Wilkinson, D. (1993) 'Maastricht and the environment: the implications for the EC's environment policy', *Journal of Environmental Law*, 4, 2: 221–39.

Winter, M. (1996) *Rural Politics: Politics for Agriculture, Forestry and the Environment*. London: Routledge.

World Wide Fund for Nature *et. al.* (1991) *Greening the Treaty*, Brussels: WWF European Policy Office.

—— (1994) *MADE in Europe: Models of Alternative Development in Europe – The Sustainable Development Imperative and the Role of the European Commission*, Brussels: WWF European Policy Office.

Wynne, B. (1987) *Risk Management and Hazardous Wastes: Implementation and the Dialectics of Credibility*, London/Berlin/New York: Springer Verlag.

—— (1996) 'May the sheep safely graze?', in Lash, S., Szerszynski, B. and Wynne, B. (eds) *Risk, Environment and Modernity: Towards a New Ecology*, London: Sage.

Wynne, G.R., Avery, M.I., Campbell, L., Gubbay, S., Hawkswell, S., Juniper, A., King, M., Newbery, P., Smart, J., Steel, C., Stones, A., Stubbs, A., Taylor, J.P., Tydeman, C. and Wynde, R. (1994) *Biodiversity Challenge: An Agenda for Conservation in the UK* (2nd edn), Sandy: Royal Society for the Protection of Birds.

Zaide Pritchard, S. (1987) *Oil Pollution Control*, London: Croom Helm.

INDEX